CLIMATE CHANGE AND ITS CAUSES, EFFECTS AND PREDICTION

CLIMATE CHANGE ADAPTATION: STEPS FOR A VULNERABLE PLANET (WITH DVD)

CLIMATE CHANGE AND ITS CAUSES, EFFECTS AND PREDICTION

Additional books in this series can be found on Nova's website under the Series tab.

Additional E-books in this series can be found on Nova's website under the E-books tab.

CLIMATE CHANGE AND ITS CAUSES, EFFECTS AND PREDICTION

CLIMATE CHANGE ADAPTATION: STEPS FOR A VULNERABLE PLANET (WITH DVD)

ELIZABETH N. BREWSTER
EDITOR

Nova Science Publishers, Inc.
New York

Copyright © 2010 by Nova Science Publishers, Inc.

All rights reserved. No part of this book may be reproduced, stored in a retrieval system or transmitted in any form or by any means: electronic, electrostatic, magnetic, tape, mechanical photocopying, recording or otherwise without the written permission of the Publisher.

For permission to use material from this book please contact us:
Telephone 631-231-7269; Fax 631-231-8175
Web Site: http://www.novapublishers.com

NOTICE TO THE READER

The Publisher has taken reasonable care in the preparation of this book, but makes no expressed or implied warranty of any kind and assumes no responsibility for any errors or omissions. No liability is assumed for incidental or consequential damages in connection with or arising out of information contained in this book. The Publisher shall not be liable for any special, consequential, or exemplary damages resulting, in whole or in part, from the readers' use of, or reliance upon, this material. Any parts of this book based on government reports are so indicated and copyright is claimed for those parts to the extent applicable to compilations of such works.

Independent verification should be sought for any data, advice or recommendations contained in this book. In addition, no responsibility is assumed by the publisher for any injury and/or damage to persons or property arising from any methods, products, instructions, ideas or otherwise contained in this publication.

This publication is designed to provide accurate and authoritative information with regard to the subject matter covered herein. It is sold with the clear understanding that the Publisher is not engaged in rendering legal or any other professional services. If legal or any other expert assistance is required, the services of a competent person should be sought. FROM A DECLARATION OF PARTICIPANTS JOINTLY ADOPTED BY A COMMITTEE OF THE AMERICAN BAR ASSOCIATION AND A COMMITTEE OF PUBLISHERS.

Additional color graphics may be available in the e-book version of this book.

LIBRARY OF CONGRESS CATALOGING-IN-PUBLICATION DATA
Climate change adaptation : steps for a vulnerable planet / editors,
Elizabeth N. Brewster.
  p. cm.
Includes index.
ISBN 978-1-61728-889-0 (hardcover)
1. Climatic changes--Government policy--United States. 2. Climatic changes--United States. 3. Global warming--Government policy--United States. 4. Global warming--United States. I. Brewster, Elizabeth N.
QC903.2.U6C56 2010
363.738'74--dc22
                       2010029855

Published by Nova Science Publishers, Inc. † New York

CONTENTS

Preface		vii
Chapter 1	Climate Change Adaptation: Strategic Federal Planning Could Help Government Officials Make More Informed Decisions *United States Government Accountability Office*	1
Chapter 2	Opening Statement of Edward J. Markey, before the Select Committee on Energy Independence and Global Warming, Hearing on "Building U.S. Resilience to Global Warming Impacts"	79
Chapter 3	Testimony of John B. Stephenson, Director, Natural Resources and Environment, before the Select Committee onEnergy Independence and Global Warming	81
Chapter 4	Testimony of Eric Schwaab, Deputy Secretary Maryland Department of Natural Resources before The U.S. House of Representatives Select Committee on Energy Independence and Global Warming	89

Chapter 5	Testimony of Stephen Seidel, Vice President for Policy Analysis, Pew Center on Global Climate Change, before the Select Committee on Energy Independence and Global Warming, Hearing on "The Federal Government's Role in Building Resilience to Climate Change"	**97**
Chapter 6	Testimony of Dr. Kenneth P. Green Before the House Select Committee on Energy Independence and Global Warming Hearing "Building U.S. Resilience to Global Warming Impacts"	**109**
Chapter 7	Comparison of Climate Change Adaptation Provisions in S. 1733 and H.R. 2454 *Melissa D. Ho*	**113**

Chapter Sources   **157**

Index   **159**

Included on Accompanying DVD:

Preliminary Review of Adaptation Options
for Climate-Sensitive Ecosystems and Resources
United States Climate Change Science Program

PREFACE

Changes in the climate attributable to increased concentrations of greenhouse gases may have significant impacts in the United States and the world. For example, climate change could threaten coastal areas with rising sea levels. Greenhouse gases already in the atmosphere will continue altering the climate system into the future, regardless of emissions control efforts. Therefore, adaptation, defined as adjustments to natural or human systems in response to actual or expected climate change, is an important part of the response to climate change. This book examines the complex issue of climate change and suggests that government-wide strategic planning, with the commitment of top leaders, can integrate activities that span a wide array of federal, state, and local entities.

Chapter 1- Changes in the climate attributable to increased concentrations of greenhouse gases may have significant impacts in the United States and the world. For example, climate change could threaten coastal areas with rising sea levels. Greenhouse gases already in the atmosphere will continue altering the climate system into the future, regardless of emissions control efforts. Therefore, adaptation—defined as adjustments to natural or human systems in response to actual or expected climate change—is an important part of the response to climate change.

GAO was asked to examine (1) what actions federal, state, local, and international authorities are taking to adapt to a changing climate; (2) the challenges that federal, state, and local officials face in their efforts to adapt; and (3) actions that Congress and federal agencies could take to help address these challenges. We also discuss our prior work on similarly complex, interdisciplinary issues. This report is based on analysis of studies, site visits to areas pursuing adaptation efforts, and responses to a Web-based questionnaire sent to federal, state, and local officials.

Chapter 2- We all remember the tragic consequences of Hurricane Katrina – the breached levees, water-filled streets, and families seeking shelter in the Superdome. While many individuals courageously responded to this disaster, government leadership failed the people of New Orleans when they needed help most. Katrina foreshadows the consequences of climate change if we do not make the necessary preparations.

Since then, scientists have shown that the warming of our climate system from emissions of heat-trapping gases – from our tailpipes and smokestacks – is unequivocal.

We face not only an increasing number of strong storms, but also many permanent alterations that will affect people throughout the country. Coastal cities like Boston will be at risk of inundation from sea level rise, which is accelerating as our oceans warm and our polar ice caps melt. Alaskan villages are finding the land they call home literally melting out from underneath them as the permafrost thaws. In the West, our shrinking mountain snowpack strains our water resource systems. Throughout this country, our farms are threatened by rising temperatures, water scarcity, and pests. For a projected 2.2 degree (Fahrenheit) rise in temperatures over the next 30 years, we can expect significant declines in the crops that make up the base of our food system.

Chapter 3- I am pleased to be here today to discuss our report to this committee on climate change adaptation and the role strategic federal planning could play in government decision making. Changes in the climate attributable to increased concentrations of greenhouse gases may have significant impacts in the United States and internationally.[1] For example, climate change could threaten coastal areas with rising sea levels. In recent years, climate change adaptation—adjustments to natural or human systems in response to actual or expected climate change—has begun to receive more attention because the greenhouse gases already in the atmosphere are expected to continue altering the climate system into the future, regardless of efforts to control emissions. According to a recent report by the National Research Council (NRC), however, individuals and institutions whose futures will be affected by climate change are unprepared both conceptually and practically for meeting the challenges and opportunities it presents. In this context, adapting to climate change requires making policy and management decisions that cut across traditional economic sectors, jurisdictional boundaries, and levels of government. My testimony is based on our October 2009 report,[2] which is being publicly released today, and addresses three issues: (1) what actions federal, state, local, and international authorities are taking to adapt to a changing climate; (2) the challenges that federal, state, and local officials face in their efforts to adapt; and (3) the actions that Congress and federal agencies could take to help address these challenges. We also provide information about our prior work on similarly complex, interdisciplinary issues

Chapter 4- Given our more than 4,000 miles of coastline and documented rate of sea level rise nearly twice that of the global average, Maryland has already begun to strategically plan for the impacts of climate change. In April 2007, Governor Martin O'Malley signed an Executive Order establishing the Maryland Climate Change Commission. Approximately a year after its formation, the Commission released Maryland's Climate Action Plan', setting forth a course of action to stem not only the drivers of climate change but also for how to adapt and respond to the inevitable consequences.

Historic tide-gauge records show that sea levels are rising along Maryland's coast and have increased one-foot within state waters over the last 100 years. We are currently expecting that sea level may rise at least twice as fast as it did during the 20th century, resulting in potentially 2.7 to 3.4 feet of rise by the year 2100. Such a rise will likely cause increased vulnerability to storm events, more frequent and severe coastal flooding, inundation of low-lying lands, submergence of tidal marshes, more shore erosion, salt-water intrusion, and higher water tables. While Maryland's entire coast will be impacted over the course of time, our state's low- lying coastal areas, to as well as those with large amounts of exposed shoreline are most at risk. The Chesapeake Bay is ranked the third most vulnerable region in the nation to the impact of sea level rise.

Chapter 5- Responding to the risks of climate change represents one of the major challenges facing our nation and the global community. Most of the attention to date has appropriately been placed on actions to reduce emissions of greenhouse gases. This is obviously the first and best line of defense against the risks associated with global warming. But as our scientific understanding of climate change has improved, we also have come to realize that our past emissions have already begun to affect our current climate. Climate change isn't some distant concern that will impact our children or grandchildren. There is clear and convincing evidence that we have already experienced the following changes:

Chapter 6- My biases are for solving environmental problems, whenever possible, with instruments that maximize freedom, opportunity, enterprise, and personal responsibility. Thus, I strongly favor true market-based remedies for environmental problems over command-and-control regulation. (I will observe here that cap-and-trade legislation is not truly market-based, as government sets a limit on emissions, rather than allowing a market to determine that level. Cap-and-trade is more akin to rationing than it is to markets).

Finally, my beliefs are based on reading the scientific literature as well as the IPCC climate science reports, and I believe that while greenhouse gases do retain heat in the atmosphere (making Earth habitable), the heat-retention capability of additional anthropogenic greenhouse gases is modest. I do not believe in predictive climate models, or most other forms of forecasting other than simple extrapolation for very modest periods of time.

Chapter 7- This report summarizes and compares climate change adaptation-related provisions in the American Clean Energy and Security Act of 2009 (H.R. 2454) and the Clean Energy, Jobs, and Power Act (S. 1733). H.R. 2454 was introduced by Representatives Waxman and Markey and passed the House on June 26, 2009. S. 1733 was introduced to the Senate by Senators Boxer and Kerry and, after subsequent revisions made in the form of a manager's substitution amendment, was reported out of the Senate Environment and Public Works Committee on November 5, 2009.

Adaptation measures aim to improve an individual's or institution's ability to cope with or avoid harmful impacts of climate change, and to take advantage of potential beneficial ones. Both H.R. 2454 and S. 1733 include adaptation provisions that (1) seek to better assess the impacts of climate change and variability that are occurring now and in the future; and (2) support adaptation activities related to climate change, both domestically and internationally.

Overall, while the two bills would authorize similar adaptation programs, they differ somewhat in scope and emphasis, and they also differ in the distribution of emission allowance allocations over time. Both bills contain provisions that address international climate change adaptation; domestic climate change adaptation programs, including the U.S. Global Change Research Program (USGCRP), the National Climate Service, and state and tribal programs; public health; and natural resources adaptation. S. 1733 includes five additional provisions not provided for in the House bill that deal with drinking water utilities; water system mitigation and adaptation partnerships; flood control, protection, prevention, and response; wildfire; and coastal Great Lakes states' adaptation.

In: Climate Change Adaptation
Editor: Elizabeth N. Brewster

ISBN: 978-1-61728-889-0
© 2010 Nova Science Publishers, Inc.

Chapter 1

CLIMATE CHANGE ADAPTATION: STRATEGIC FEDERAL PLANNING COULD HELP GOVERNMENT OFFICIALS MAKE MORE INFORMED DECISIONS

United States Government Accountability Office

WHY GAO DID THIS STUDY

Changes in the climate attributable to increased concentrations of greenhouse gases may have significant impacts in the United States and the world. For example, climate change could threaten coastal areas with rising sea levels. Greenhouse gases already in the atmosphere will continue altering the climate system into the future, regardless of emissions control efforts. Therefore, adaptation—defined as adjustments to natural or human systems in response to actual or expected climate change—is an important part of the response to climate change.

GAO was asked to examine (1) what actions federal, state, local, and international authorities are taking to adapt to a changing climate; (2) the challenges that federal, state, and local officials face in their efforts to adapt; and (3) actions that Congress and federal agencies could take to help address these challenges. We also discuss our prior work on similarly complex, interdisciplinary issues. This report is based on analysis of studies, site visits to areas pursuing adaptation efforts, and responses to a Web-based questionnaire sent to federal, state, and local officials.

WHAT GAO RECOMMENDS

ds GAO recommends that within the Executive Office of the President the appropriate entities, such as the Council on Environmental Quality (CEQ), develop a national adaptation plan that includes setting priorities for federal, state, and local agencies. CEQ generally agreed with our recommendations.

WHAT GAO FOUND

While available information indicates that many governments have not yet begun to adapt to climate change, some federal, state, local, and international authorities have started to act. For example, the U.S. National Oceanic and Atmospheric Administration's Regional Integrated Sciences and Assessments program supports research to meet the adaptation-related information needs of local decision makers. In another example, the state of Maryland's strategy for reducing vulnerability to climate change focuses on protecting habitat and infrastructure from future risks associated with sea level rise and coastal storms. Other GAO discussions with officials from New York City; King County, Washington; and the United Kingdom show how some governments have started to adapt to current and projected impacts in their jurisdictions.

The challenges faced by federal, state, and local officials in their efforts to adapt fell into three categories, based on GAO's analysis of questionnaire results, site visits, and available studies. First, competing priorities make it difficult to pursue adaptation efforts when there may be more immediate needs for attention and resources. For example, about 71 percent (128 of 180) of the officials who responded to our questionnaire rated "non-adaptation activities are higher priorities" as very or extremely challenging. Second, a lack of site-specific data, such as local projections of expected changes, can reduce the ability of officials to manage the effects of climate change. For example, King County officials noted that they are not sure how to translate climate data into effects on salmon recovery. Third, adaptation efforts are constrained by a lack of clear roles and responsibilities among federal, state, and local agencies. Of particular note, about 70 percent (124 of 178) of the respondents rated the "lack of clear roles and responsibilities for addressing adaptation across all levels of government" as very or extremely challenging.

GAO's analysis also found that potential federal actions for addressing challenges to adaptation efforts fell into three areas. First, training and education efforts could increase awareness among government officials and the public about the impacts of climate change and available adaptation strategies. Second, actions to provide and interpret site-specific information would help officials understand the impacts of climate change at a scale that would enable them to respond. For instance, about 80 percent (147 of 183) of the respondents rated the "development of state and local climate change impact and vulnerability assessments" as very or extremely useful. Third, Congress and federal agencies could encourage adaptation by clarifying roles and responsibilities. About 71 percent (129 of 181) of the respondents rated the development of a national adaptation strategy as very or extremely useful.

Climate change is a complex, interdisciplinary issue with the potential to affect every sector and level of government operations. Our past work on crosscutting issues suggests that governmentwide strategic planning—with the commitment of top leaders—can integrate activities that span a wide array of federal, state, and local entities.

October 7, 2009
The Honorable Edward Markey
Chairman
Select Committee on Energy Independence and Global Warming
House of Representatives
Dear Mr. Chairman:

Changes in the earth's climate attributable to increased concentrations of greenhouse gases may have significant environmental and economic impacts in the United States and internationally.[1] Among other potential impacts, climate change could threaten coastal areas with rising sea levels, alter agricultural productivity, and increase the intensity and frequency of floods and tropical storms. Federal, state, and local agencies are tasked with a wide array of responsibilities, such as managing natural resources, that will be affected by a changing climate. Furthermore, climate change has implications for the fiscal health of the federal government, affecting federal crop and flood insurance programs, and placing new stresses on infrastructure. The effects of increases in atmospheric concentrations of greenhouse gases and temperature on ecosystems are expected to vary across regions (see table 1).

Proposed responses to climate change include reducing greenhouse gas emissions through regulation, promoting low-emissions technologies, and adapting to the possible impacts by planning and improving protective infrastructure. Thus far, federal government attention and resources have been focused on emissions reduction options, climate science research, and technology investment. In recent years, however, climate change adaptation—adjustments to natural or human systems in response to actual or expected climate change—has begun to receive more attention because the greenhouse gases already in the atmosphere are expected to continue altering the climate system into the future, regardless of efforts to control emissions.

Policymakers are increasingly viewing adaptation as a risk-management strategy to protect vulnerable sectors and communities that might be affected by changes in the climate. As the Director of the Office of Science and Technology Policy in the Executive Office of the President stated in a 2009 testimony, we can invest in countless ways to reduce our vulnerability to the changes in climate that we do not succeed in avoiding, for example by breeding heat- and drought-resistant crop strains, bolstering our defenses against tropical diseases, improving the efficiency of our water use, and starting to manage our coastal zones with sea level rise in mind.[2] Furthermore, certain natural resource adaptation activities— such as efforts to build large, connected landscapes—will become more important as native species attempt to migrate or otherwise adapt to climate change. While it may be costly to raise river or coastal dikes to protect communities and resources from sea level rise, build higher bridges, or improve storm water systems, there is a growing recognition, in the United States and elsewhere, that the cost of inaction could be greater.

According to a recent report by the National Research Council (NRC), however, individuals and institutions whose futures will be affected by climate change are unprepared both conceptually and practically for meeting the challenges and opportunities it presents. Many usual practices and decision rules (for building bridges, implementing zoning rules, using private motor vehicles, and so on) assume a stationary climate—a continuation of past climate conditions, including similar patterns of variation and the same probabilities of extreme events. According to NRC, that assumption, fundamental to the ways people and organizations make their choices, is no longer valid.

Table 1. Current and Projected Impacts of Climate Change in the United States.

Category	Current and projected impacts
Temperature	• U.S. average temperature has risen more than 2 degrees Fahrenheit over the past 50 years and is projected to rise more in the future—how much more depends primarily on the amount of heat-trapping gases emitted globally and how sensitive the climate is to those emissions.
Precipitation	• Precipitation has increased an average of about 5 percent over the past 50 years. Projections of future precipitation generally indicate that northern areas will become wetter and southern areas, particularly in the West, will become drier.
	• The amount of rain falling in the heaviest downpours has increased approximately 20 percent on average in the past century, and this trend is very likely to continue, with the largest increases in the wettest places.
Extreme weather events	• Many types of extreme weather events, such as heat waves and regional droughts, have become more frequent and intense during the past 40 to 50 years.
Storms	• The destructive energy of Atlantic hurricanes has increased in recent decades. The intensity of these stormsis likely to increase in this century.
	• In the eastern Pacific, the strongest hurricanes have become stronger since the 1980s, even while the total number of storms has decreased.
	• Cold season storm tracks are shifting northward, and the strongest storms are likely to become stronger andmore frequent.
Sea levels	• Sea level has risen along most of the U.S. coast over the last 50 years and will likely rise more in the future.
	• Arctic sea ice is declining rapidly and this decline is very likely to continue.

Source: Adapted from the U.S. Global Change Research Program, Global Climate Change Impacts in the United States, 2009.

Adapting to climate change requires making policy and management decisions that cut across traditional economic sectors, agencies, jurisdictional boundaries, and levels of government. The authorities and expertise necessary to facilitate adaptation activities are spread among many agencies. Recent proposed legislation considers governmentwide adaptation strategies, including the development of a National Climate Service to inform the public through the sustained production and delivery of authoritative, timely, and useful information about the impacts of climate change on local, state, regional, tribal, national, and global scales.[3] For example, the American Clean Energy and Security Act of 2009, which passed the House of Representatives on June 26, 2009, contains provisions related to climate change adaptation, including the development of federal and state natural resource agency adaptation plans and the establishment of a natural resources climate change adaptation fund.

In this context, our review (1) determines what actions, if any, federal, state, local, and international authorities are taking to adapt to a changing climate; (2) identifies the challenges, if any, that federal, state, and local officials reported facing in their efforts to adapt; and (3) identifies actions that Congress and federal agencies could take to help address these challenges. We also provide information about our prior work on similarly complex, interdisciplinary issues.

To determine the actions federal, state, local, and international authorities are taking to adapt to a changing climate, we obtained summaries of adaptation-related efforts from a broad range of federal agencies and visited four sites where government officials are taking actions to adapt.[4] We chose these sites because they were frequently mentioned in the background literature and scoping interviews as examples of locations that are implementing climate change adaptation and which may offer particularly useful insights into the types of actions governments can take to plan for climate change impacts. The four sites were New York City; King County, Washington; the state of Maryland; and the United Kingdom. Our selected sites are not representative of all adaptation efforts taking place; however, they include a variety of responses to climate change effects across different levels of government. We included an international site visit to examine how other countries are also starting to adapt. We gathered information during and after site visits through observation of adaptation efforts, interviews with officials and stakeholders, and a review of documents provided by these officials.

To describe challenges that federal, state, and local officials face in their efforts to adapt and the actions that Congress and federal agencies could take to help address these challenges, we reviewed available studies and asked knowledgeable stakeholders about challenges that federal, state, and local officials may face in adaptation efforts. Using this information, we compiled lists of potential challenges and potential actions the federal government could take to address them and developed a Web-based questionnaire to gather officials' views on these challenges and actions. We designed the questionnaire to collect aggregate information through a range of closed-ended questions, as well as illustrative examples through open-ended responses. Within the questionnaire, we organized questions about challenges and actions into groups related to the following: (1) awareness and priorities, (2) information, and (3) the structure and operation of the federal government. We worked with organizations that represent federal, state, and local officials to select a nonprobability sample of 274 officials knowledgeable about adaptation, of which 187 completed the questionnaire, for a response rate of approximately 68 percent.[5] The federal, state, and local officials who responded represent a diverse array of disciplines, including planners, scientists, and public health professionals. A more detailed description of our scope and methodology is available in appendix I.

We conducted this performance audit from September 2008 to October 2009 in accordance with generally accepted government auditing standards. Those standards require that we plan and perform the audit to obtain sufficient, appropriate evidence to provide a reasonable basis for our findings and conclusions based on our audit objectives. We believe that the evidence obtained provides a reasonable basis for our findings and conclusions based on our audit objectives.

FEDERAL, STATE, LOCAL, AND INTERNATIONAL EFFORTS TO ADAPT TO CLIMATE CHANGE

While federal agencies are beginning to recognize the need to adapt to climate change, there is a general lack of strategic coordination across agencies, and most efforts to adapt to potential climate change impacts are preliminary. However, some states and localities have begun to make progress on adaptation independently and through partnerships with other entities, such as academic institutions. The subjects of our site visits in the United States— New York City; King County, Washington; and Maryland— have all taken steps to plan for climate change and have begun to implement adaptive measures in sectors such as natural resource management and infrastructure. Their on-the-ground experiences can help inform the federal approach to adaptation, which is now primarily focused on assessing projected climate impacts and exploring adaptation options. In addition, certain nations have taken action to adapt to climate change. Our detailed examination of the United Kingdom provides an example of a country where central and local government entities are working together to address climate change impacts.

Many Federal Agencies are Beginning to Take Steps to Adapt to Climate Change

Although there is no coordinated national approach to adaptation, several federal agencies report that they have begun to take action with current and planned adaptation activities. These activities are largely ad hoc and fall into several categories, including (1) information for decision making, (2) federal land and natural resource management, (3) infrastructure design and operation, (4) public health research, (5) national security preparation, (6) international assistance to developing countries, and (7) governmentwide adaptation strategies. We provide information on selected federal efforts to adapt to climate change, submitted to us by federal agencies, in a supplement to this report (see GAO-10-114SP).

Information for decision making: A range of preliminary adaptation-related activities are reported to be under way at different agencies, including efforts to provide relevant climate information to help decision makers plan for future climate impacts. For example, two programs managed by the National Oceanic and Atmospheric Administration (NOAA) help policymakers and managers obtain the information they need to adapt to a changing climate. NOAA's Regional Integrated Sciences and Assessments (RISA) program supports climate change research to meet the needs of decision makers and policy planners at the national, regional, and local levels. Similarly, NOAA's Sectoral Applications Research Program is designed to help decision makers in different sectors, such as coastal resource managers, use climate information to respond to and plan for climate variability and change, among other goals.

Other agencies—including the National Science Foundation, the Department of the Interior (Interior), the Environmental Protection Agency (EPA), the National Aeronautics and Space Administration (NASA), and the

Department of Energy—also manage programs to provide climate information to decision makers. For example, the National Science Foundation supports the scientific research needed to help authorities and the public plan adaptation activities and address any challenges that arise. Similarly, Interior's newly formed Energy and Climate Change Task Force is working to ensure that climate change impact data collection and analysis are better integrated and disseminated, that data gaps are identified and filled, and that the translation of science into adaptive management techniques is geared to the needs of land, water, and wildlife managers as they develop adaptation strategies in response to climate change-induced impacts on landscapes. Another example of information sharing is EPA's Climate Ready Estuaries program, which provides a toolkit to coastal communities and participants in its National Estuary Program on how to monitor climate change and where to find data. In addition, NASA's Applied Sciences Program is working in 31 states and with a number of federal agencies to help officials use NASA's climate data to make adaptation decisions. For example, NASA forecasts stream temperatures for NOAA managers responsible for managing chinook salmon populations in the Sacramento River and predicts water flow regimes and subsequent fire risk in Yosemite National Park. DOE's Integrated Assessment Research Program supports research on models and tools for integrated analysis of both the drivers and consequences of climate change. DOE's supercomputing resources provide the capability to assess impacts and vulnerabilities to temperature change, anticipate extreme events, and predict risk from climate change effects (e.g., water availability) on a regional and local basis to better inform decision makers.

Federal land and natural resource management: Several federal agencies have reported beginning to consider measures that would strengthen the resilience of natural resources in the face of climate change. For example, on September 14, 2009, Interior issued an order designed to address the impacts of climate change on the nation's water, land, and other natural and cultural resources.[6] The Interior order, among other things, designated eight regional Climate Change Response Centers. According to Interior, these centers will synthesize existing climate change impact data and management strategies, help resource managers put them into action on the ground, and engage the public through education initiatives. Similarly, several federal agencies recently released draft reports required by Presidential Executive Order that describe strategies for protecting and restoring the Chesapeake Bay, including addressing the impacts of climate change on the bay.[7] In addition, the U.S.

Forest Service reported that it devotes about $9 million to adaptation research and has developed a strategic framework that recognizes the need to enhance the capacity of forests and grasslands to adapt. The Chief of the Forest Service recently testified that dealing with climate change risks and uncertainties will need to be a more prominent part of the Forest Service's management decision processes.[8]

Certain agencies have also identified specific adaptation strategies and tools for natural resource managers. For example, Interior provided a number of adaptation-related policy options for land managers in reports produced for its Climate Change Task Force, a past effort that has since been expanded upon to reflect new priorities.[9] Similarly, a recent U.S. Climate Change Science Program report provided a preliminary review of adaptation options for climate-sensitive ecosystems and resources on federally owned and managed lands.[10] In addition, the Department of Defense's Legacy Resource Management Program is working with other agencies to develop a guidance manual that will summarize available natural resource vulnerability assessment tools.

In some instances, federal agencies have begun to help implement adaptation actions. A recent Congressional Research Service presentation highlighted two case studies on federal lands in which federal agencies assisted with adaptation efforts. The first is a habitat restoration project supported by the U.S. Fish and Wildlife Service (FWS) to adapt to sea level rise in the Albemarle Peninsula, North Carolina. The second focuses on increasing landscape diversity and managing biodiversity in Washington's Olympic National Forest, the site of a Forest Service Pacific Northwest Research Station. The project involved work with the Federal Highway Administration to protect watersheds and roads.[11] In addition, the Department of Energy reported that it has assessed major water availability issues related to energy production and use, such as electrical generation and fuels production, and identified approaches that could reduce freshwater use in the energy sector, and opportunities for further research and development to address questions that decision makers will need to resolve to effectively manage the energy and water availability issues.

Infrastructure design and operation: A number of federal agencies are beginning to recognize that they must account for climate change impacts when building and repairing man-made infrastructure, since such impacts have implications beyond the natural environment.[12] Many adaptation efforts related to infrastructure are at the planning stages to date. For example, the U.S. Army Corps of Engineers' adaptation initiatives include leading a team of water managers to evaluate how climate change considerations can be incorporated into activities related to water resources. These managers are also participating in an interagency group (Climate Change and Water Working Group) which held workshops in California in spring 2007. At these workshops, water managers from federal (U.S. Geological Survey (USGS), Bureau of Reclamation, NOAA), state, local, and private agencies and organizations recommended more flexible reservoir operations, better use of forecasts, and more monitoring of real-time conditions in the watersheds. A draft report of long-term needs identified by the team was undergoing agency review in August and September 2009. In addition, EPA recently issued a guide entitled *Smart Growth for Coastal and Waterfront Communities* to help communities address challenges such as potential sea level rise and other climate-related hazards.[13]

Within the U.S. Department of Transportation (DOT), the Federal Highway Administration also formed a multidisciplinary internal working group to coordinate infrastructure policy and program activities, specifically to address climate change effects on transportation. Both the U.S. Army Corps of Engineers and DOT are reviewing the impacts of sea level rise on infrastructure. DOT found that a 2-foot sea level rise would affect 64 percent of the Gulf Coast's port facilities, while a 4-foot rise would affect nearly three-quarters of port facilities.[14] In addition, the Federal Emergency Management Agency (FEMA), part of the U.S. Department of Homeland Security, is currently conducting a study on the impact of climate change on the National Flood Insurance Program, as we recommended in a 2007 GAO report.[15] The Department of Energy is also working to protect critical infrastructure—such as the national laboratories and the Strategic Petroleum Reserve—by using climate impact assessments and developing guidance for management decisions that account for climate change.

Public health research: Federal agencies responsible for public health matters are starting to support modeling and research efforts to assess climate change impacts on their programs and issue areas. Currently, the Centers for Disease Control and Prevention's (CDC) Climate Change program is engaged in a number of adaptation initiatives that address various populations' vulnerability to the adverse health effects of heat waves. For example, CDC helped develop a Web-based modeling tool to assist local and regional governments to prepare for heat waves and an extreme heat media toolkit for cities.

In addition, the National Institutes of Health (NIH) formed a working group on Climate Change and Health, which aims to identify research needs and priorities and involve the biomedical research community in discussions of the health effects of climate change. Recently, NIH developed an initiative called the NIH Challenge Grants in Health and Science Research, which supports research on predictive climate change models and facilitates public health planning. Of particular interest to NIH are studies that quantify the current impacts of climate on a variety of communicable or noncommunicable diseases or studies that project the impacts of different climate and socioeconomic scenarios on health. EPA is also taking steps to ensure that public health needs are met in the context of climate change. For example, EPA helped produce an analysis that examined potential impacts of climate change on human society, opportunities for adaptation, and associated recommendations for addressing data gaps and research goals.[16] In addition, EPA is working with agencies such as CDC, NIH, and NOAA to support the public health communities' efforts to develop strategies for adapting to climate change.

National security preparation: Federal agencies are beginning to study the potential consequences of climate change on national security. For example, the Department of Defense's ongoing Quadrennial Defense Review is examining the capabilities of the armed forces to respond to the consequences of climate change—in particular, preparedness for natural disasters from extreme weather events, as is required by Section 951 of the National Defense Authorization Act for fiscal year 2008.[17] This act also requires the department to develop guidance for military planners to assess the risk of projected climate change, update defense plans based on these assessments, and develop the capabilities needed to reduce future impacts. In October 2008, the Air Force participated in a Colloquium on National Security Implications of Climate Change sponsored by the U.S. Joint Forces Command. In addition, the Navy recently sponsored a Naval Studies Board study on the National Security Implications of Climate Change on U.S. Naval forces (Navy, Marine Corps, and Coast Guard), to be completed in late 2010. This study is intended to help the Navy develop future robust climate change adaptation strategies.

International assistance to developing countries: Some federal agencies are supporting preliminary adaptation planning efforts internationally. For example, the U.S. Agency for International Development (USAID) funds climate change activities related to agriculture, water, forest, and coastal zone management in partner developing countries. To inform such activities, USAID produced two documents, an adaptation guidance manual and a coastal zone adaptation manual, which provide climate change tools and other information to planners in the developing world.[18] In addition, USAID works with NASA to provide developing countries with climate change data to help support adaptation activities. For example, the two agencies use SERVIR, a high-tech regional satellite visualization and monitoring system for Central America, to provide a climate change scenario database, climate change maps indicating impacts on Central America's biodiversity, a fire and smoke mapping and warning system, red tide alerts, and weather alerts. The U.S. Department of State's and NOAA's climate efforts also sustain adaptation initiatives worldwide. NOAA is supporting USAID programs in Asia, Latin America, and Africa by using a science-based approach to enhance governments' abilities to understand, anticipate, and manage climate risk. In addition, Interior's International Technical Assistance Program, funded through interagency agreements with USAID and the U.S. Department of State, provides training and technical assistance to developing countries.[19]

Governmentwide adaptation strategies: Currently, no single entity is coordinating climate change adaptation efforts across the federal government and there is a general lack of strategic coordination. However, several federal entities are beginning to develop governmentwide strategies to adapt to climate change. For example, the President's Council on Environmental Quality (CEQ) is leading a new initiative to coordinate the federal response to climate change in conjunction with the Office of Science and Technology Policy, NOAA, and other agencies. Similarly, the U.S. Global Change Research Program (USGCRP), which coordinates and integrates federal research on climate change, has developed a series of "building blocks" that outline options for future climate change work, including science to inform adaptation. The adaptation building block includes support and guidance for federal, regional, and local efforts to prepare for and respond to climate change, including characterizing the need for adaptation and developing, implementing, and evaluating adaptation approaches.

Certain State and Local Governments Are Developing and Implementing Climate Change Adaptation Measures

Many government authorities at the state and local levels have not yet begun to adapt to climate change. According to a recent NRC report, the response of governments at all levels, businesses and industries, and civil society is only starting, and much is still to be learned about the institutional, technological, and economic shifts that have begun.[20] Some states have not yet started to consider mitigation or adaptation; others have developed plans but have not yet begun to implement them. However, certain governments are beginning to plan for the effects of climate change and to implement climate change adaptation measures. For example, California recently issued a draft climate adaptation strategy, which directs the state government to prepare for rising sea levels, increased wildfires, and other expected changes.[21] A general review of state and local government adaptation planning efforts is available in two recent reports issued by nongovernment research groups.[22]

We visited three U.S. sites—New York City; King County, Washington; and the state of Maryland—where government officials have begun to plan for and respond to climate change impacts. The three locations are all addressing climate change adaptation to various extents. New York City is in the planning phases for its citywide efforts, although individual departments have begun to implement specific actions, such as purchasing land in New York City's watershed to improve the quality of its water supply. King County, Washington has, among other things, completed and begun to implement a comprehensive climate change plan, which includes an adaptation component. Maryland has released the first phase of its adaptation strategy, which is focused on sea level rise and coastal storms, reflecting sectors of immediate concern.

Our analysis of these sites suggests three major factors have led these governments to act. First, natural disasters such as floods, heat waves, droughts, or hurricanes raised public awareness of the costs of potential climate change impacts. Second, leaders in all three sites used legislation, executive orders, local ordinances, or action plans to focus attention and resources on climate change adaptation. Finally, each of the governments had access to relevant site-specific information to provide a basis for planning and management efforts. This site-specific information arose from partnerships that decision makers at all three sites formed with local universities and other government and nongovernment entities.

The following summaries describe the key factors that motivated these governments to act, the policies and laws that guide adaptation activities at each location, the programs and initiatives that are in place to address climate effects, the sources of site-specific information, and any partnerships that have assisted with adaptation activities.

New York City, New York

New York City's adaptation efforts stemmed from a growing recognition of the vulnerability of the city's infrastructure to natural disasters, such as the severe flooding in 2007 that led to widespread subway closures. The development of PlaNYC—a plan to accommodate a projected population growth of 1 million people, reduce citywide carbon emissions by 30 percent, and make New York City a greener, more sustainable city by 2030—also pushed city officials to think about the future, including the need for climate change adaptation. New York City's extensive coastline and dense urban infrastructure makes it vulnerable to sea level rise; flooding; and other extreme weather, including heatwaves, which could become more common as a result of climate change.

City officials took several steps to formalize a response to climate change. In 2008, the Mayor convened the New York City Panel on Climate Change (NPCC) to provide localized climate change projections and decision tools. The Mayor also invited public agencies and private companies to be part of the New York City Climate Change Adaptation Task Force, a public-private group charged with assessing climate effects on critical infrastructure and developing adaptation strategies to reduce these risks. The Office of Long-Term Planning and Sustainability, established by a local law in 2008, provides oversight of the city's adaptation efforts, which are part of PlaNYC.[23] In addition to citywide efforts, a number of municipal and regional agencies have begun to address climate change adaptation in their operations.

To date, New York City's adaptation efforts typically have been implemented as facilities are upgraded or as funding becomes available. For example, the city's Department of Environmental Protection (DEP), which manages water and wastewater infrastructure, has begun to address flood risks to its wastewater treatment facilities. These and other efforts are described in DEP's *2008 Climate Change Program Assessment and Action Plan*.[24] Many of New York City's wastewater treatment plants, such as Tallman Island (see figure 1) are vulnerable to sea level rise and flooding from storm surges because they are located in the floodplain next to the waterbodies into which they discharge. In response to this threat, DEP is, in the course of scheduled renovations, raising sensitive electrical equipment, such as pumps and motors, to higher levels to protect them from flood damage.

Source: GAO. The Tallman Island Water Pollution Control Plant, located on the bank of the East River, is vulnerable to flooding due to storm surges and sea level rise.

Figure 1. Tallman Island Water Pollution Control Plant, Queens, New York City.

Source: GAO. Sedum (left) and native plants (right) are used in the green roof at the Five Borough Technical Services Facility.

Figure 2. New York City Department of Parks and Recreation Green Roofs at Five Borough Technical Services Facility.

Other municipal departments are implementing climate change adaptation measures as well. For example, the Department of Parks and Recreation launched a pilot project in its Five Borough Technical Services Facility to experiment with different types of green roofs—vegetated plots on rooftops that absorb rainwater and moderate the effects of heatwaves (see figure 2). According to an official at the Department of Parks and Recreation, the department plans to install green roofs in some of its recreation facilities in the next few years, since these facilities will be replacing their roofs. Green roofs are part of a suite of measures the city is exploring to control stormwater at the source (the location where the rain falls), rather than pipe it elsewhere. This can help reduce the need for more infrastructure investments in preparation for more intense rainstorms— investments that can be very costly and that are not always feasible in the space available under the city streets.

New York City's adaptation efforts have benefited from officials' access to site-specific information, starting with the publication of a 2001 report for USGCRP, which provided a scientific assessment of climate change effects in the New York City metropolitan region.[25] More recently, the city, through the financial support of the Rockefeller Foundation, created NPCC. According to its co-chairs, NPCC is charged with completing several decision-support documents, which will provide decision makers with information about local climate effects.[26] In addition, the Mayor convened the New York City Climate Change Adaptation Task Force to prepare a risk-based assessment of how climate change would affect the communication, energy, water and waste, transportation, and policy sectors, as well as the urban ecosystem and parks, and prioritize potential response strategies. Members of the task force, several of whom represent private industries, explained that they agreed to participate in the task force because the Mayor made this issue a priority. They noted that events such as Hurricane Katrina in 2005; the power outage in August 2003, which affected New York City as well as other locations in the United States and Canada; and the 2007 subway flooding raised their awareness about the effects of climate change on their operations.

New York City partners with other state and local governments to share knowledge and implement climate change adaptation efforts. It is a charter member of the C40, a coalition of large cities around the world committed to addressing climate change. City agencies also share information with counterparts in other locations about specific concerns. For example, DEP shares information about addressing water-related climate change effects with the state of California and the Water Utility Climate Alliance, a national coalition of water and wastewater utilities. DEP coordinates with other state and local governments to address climate change effects on its watershed, which is located outside of city limits. Similarly, transportation agencies that serve New York City, such as the Metropolitan Transit Authority and New Jersey Transit, cross local and state boundaries and require coordination on a regional scale, which New York City addresses through its multijurisdictional task force. City officials and members of NPCC stated that a coherent federal response would further facilitate the development of common objectives across local and state jurisdictions.

King County, Washington

According to officials from the King County Department of Natural Resources and Parks (DNRP), the county took steps to adapt to climate change because its leadership was highly aware of climate impacts on the county and championed the need to take action. The county commissioned an internal study in 2005 that included each department's projection of its operations in 2050, which focused attention on the need to prepare for future climate changes. The county also sponsored a conference in 2005 that brought together scientists, local and state officials, the private sector, and the public to discuss the impacts of climate change.[27] This conference served to educate the public and officials and spur action.

Officials from DNRP noted that recent weather events increased the urgency of certain adaptive actions. For example, in November 2006, the county experienced severe winter storms that caused a series of levees to crack. The levees had long needed repair, but the storm damage helped increase support for the establishment of a countywide flood control zone district, funded by a dedicated property tax.[28] The flood control zone district will use the funds, in part, to upgrade flood protection facilities, which will increase the county's resilience to future flooding. In addition to more severe winter storms, the county expects that climate change will lead to sea level rise; reduced snowpack; and summertime extreme weather such as heat waves and drought, which can lead to power shortages because hydropower is an important source of power in the region.

The county's first formal step toward adaptation was a climate change plan developed in 2007.[29] The county executive also issued several executive orders that call for, among other things, the evaluation of climate impacts in the State Environmental Policy Act reviews conducted by county departments and the consideration of global warming adaptation in county operations, such as transportation, waste and wastewater infrastructure, and land use planning.[30] For example, King County officials told us that during the construction of the Brightwater wastewater treatment plant, DNRP's Wastewater Treatment Division added a pipeline that could convey approximately 7 million gallons per day of reclaimed water to industrial and agricultural users upon completion in 2011. They also said that additional reclaimed water could be made available in the future as the need arises. The division is also addressing the effects of sea level rise by, for example, increasing the elevation of vulnerable facilities during design and installing flaps to prevent backflow into its pipelines. Additionally, in 2008, the county incorporated consideration of climate change into the revision of its Comprehensive Plan, which guides land use decisions throughout the county.[31]

King County officials told us that each county department convened internal teams that identify climate change initiatives and report to the King County Executive Action Group on Climate Change on their progress. For example, the county's Department of Transportation Road Services Division started a Climate Change Team in 2008, which identified several initiatives in response to projections for more intense storms, including investigating new approaches to stormwater treatment. Specifically, the Road Services Division is piloting a roadside rain garden, which captures and filters rainwater using vegetation and certain types of soil, to determine whether more of such installations could improve the onsite management of stormwater runoff, as compared to a traditional engineering approach, which would pipe the water to a pond or holding vault and then discharge it to an offsite waterbody (see figure 3). Alongside the rain garden, a permeable concrete sidewalk will absorb additional rain that would normally flow off a traditional impervious sidewalk into adjacent property. The rain garden and permeable sidewalk are considered examples of "low-impact development," which are expected to help the county adapt to increased rainfall by reducing peak surface water flows from road surfaces by about 33 percent. The Road Services Division is also implementing other measures that could improve its response to storms, such as installing larger culverts, improving its ability to detect hazardous road conditions (for example, due to flooding), and communicating those conditions to maintenance staff and the general public.

County officials receive information on climate change effects from a number of sources. The University of Washington Climate Impacts Group (CIG), funded by NOAA's RISA program, has had a long-standing relationship with county officials and works closely with them to provide regionally specific climate change data and modeling, such as a 2009 assessment of climate impacts in Washington, as well as decision-making tools.[32] For example, the CIG Web site provides a Climate Change Streamflow Scenario Tool, which allows water planners in the Columbia River basin to compare historical records with climate change scenarios. Similarly, according to its faculty, the Washington State University Extension Office works with the county and CIG to provide climate change information to the agricultural and forestry sectors, both of which will be increasingly affected by insect infestation due to increases in temperatures. The university's Extension Office also provides direct technical assistance to landowners affected by these impacts. King County officials, according to the director of DNRP, also share information about climate change adaptation with other localities through several partnership efforts, including the Center for Clean Air Policy Urban Leaders Adaptation Initiative.

Source: King County Department of Transportation Road Services Division.
This rain garden, which is under construction, treats roadway runoff using natural vegetation and certain types of soil.

Figure 3. Rain Garden in King County, Washington

Maryland

The Secretary of the Maryland Department of Natural Resources (DNR) told us that Maryland began to work on climate change adaptation because of the state's vulnerability to coastal flooding due to sea level rise and severe storms. The Maryland coastline is particularly vulnerable due to a combination of global sea level rise and local land subsidence, or sinking, among other factors. It has already experienced a sea level rise of about 1 foot in the last 100 years, which led to the disappearance of 13 Chesapeake Bay islands. According to a recent state report, a 2- to 3-foot sea level rise could submerge thousands of acres of tidal wetlands; low- lying lands; and Smith Island in the Chesapeake Bay.[33] These ongoing concerns, along with widespread flooding caused by Hurricane Isabel in 2003, have increased awareness of climate change effects in the state.

Maryland officials have taken a number of steps to formalize their response to climate change effects. An executive order in 2007 established the Maryland Commission on Climate Change, which released the Maryland Climate Action Plan in 2008.[34] As part of this effort, DNR chaired an Adaptation and Response Working Group, which issued a report on sea level rise and coastal storms.[35] The 2008 Maryland Climate Action Plan calls for future adaptation strategy development to cover other sectors such as agriculture and human health.

Maryland also enacted several legislative measures that address coastal concerns, including the Living Shoreline Protection Act of 2008, which generally requires the use of nonstructural shoreline stabilization measures instead of "hard" structures such as bulkheads and retaining walls (see figure 4).[36] According to a Maryland official, as sea level rises there will be a greater need for shore protection. Living shorelines provide such protection, while also maintaining coastal processes and providing aquatic habitat. The Chesapeake and Atlantic Coastal Bays Critical Area Protection law was also amended to, among other things, require the state to update the maps used to determine the boundary of the critical areas at least once every 12 years.[37] Previously, the critical areas were based on a map drawn in 1972 that did not reflect changes caused by sea level rise or other coastal erosion processes.

Source: GAO.
This living shoreline uses marsh plants and other natural features to protect the shore from erosion.

Figure 4. A Living Shoreline, Annapolis, Maryland

According to officials from DNR, the department is modifying several existing programs to ensure that the state is taking the effects of climate change into account. For example, an official from DNR told us that it is incorporating climate change into its ranking criteria for state land conservation. Specifically, this official said that DNR plans to prioritize coastal habitat for potential acquisition according to its suitability for marsh migration, among other factors. Additionally, Maryland is providing guidance to coastal counties to assist them with incorporating the effects of climate change into their planning documents. For example, DNR funded guidance documents to three coastal counties, Dorchester, Somerset, and Worcester Counties, on how to address sea level rise and other coastal hazards in their local ordinances and planning efforts.[38] Furthermore, in spring 2009, DNR officials participated in a public Somerset County sea level rise workshop designed to raise the awareness of county residents. Officials discussed what sea level rise projections could mean to the county, including the inundation of some of its coastal infrastructure and salt marsh habitat (see figure 5), and described some of the state initiatives to address these effects. Finally, officials with the DNR Monitoring and Non-Tidal Assessment Division told us they are considering expanding their monitoring of sentinel sites—pristine streams where changing conditions can help detect localized impacts of climate change.

Maryland draws on local universities, federal agencies, and others to access information relevant to climate change. For example, in 2008, scientists from the University of Maryland chaired and participated in the Scientific and Technical Working Group of the Maryland Commission on Climate Change. Faculty from the University of Maryland also provide technical information to the state government and legislature on an ongoing basis. Maryland receives grants and additional technical assistance from the federal government and collaborates with federal agencies and local universities to collect and disseminate data relevant to climate change adaptation. Specifically, Maryland used local, state, and federal resources to map its coastline using Light Detection and Ranging technology and has made this information, as well as a number of tools that can be used by the public and decision makers, readily available in the *Maryland Shorelines Online* Web site.[39] For example, an interactive mapping application called Shoreline Changes Online allows users to access historic shoreline data to determine erosion trends.[40]

Source: GAO.
Salt marshes in Somerset County provide important habitat to migrating waterfowl and other species; they are at risk of inundation due to sea level rise.

Figure 5. Salt Marsh in Somerset County, Maryland

Some Countries Have Begun to Adapt to Climate Change

Limited adaptation efforts are also taking root in other countries around the world. In 2007, the Intergovernmental Panel on Climate Change's Fourth Assessment Report found that some adaptation measures are being implemented in both developing and developed countries, but that many of these measures are in the preliminary stages.[41] As in the case of the state and local efforts described earlier, some of these adaptation efforts have been triggered by the recognition that current weather extremes and seasonal changes will become more frequent in the future. For example, recognizing the hazards of rising temperatures, efforts are under way in Nepal to drain the expanding Tsho Rolpa glacial lake to reduce flood risk. Similarly, in response to reduced snow cover and glacial retreat, the winter tourism industry in the European Alpine region has implemented a number of measures, such as building reservoirs to support artificial snowmaking.

A number of countries have begun to assess their vulnerability to climate change impacts and formulate national responses. For example, Canada issued a report in 2008 that discusses the current and future risks and opportunities that climate change presents, primarily from a regional perspective.[42] Australia recently issued guidance to local governments about expected climate change projections, impacts, and potential responses.[43] In addition, under the United Nations Framework Convention on Climate Change, least-developed countries can receive funding to develop National Adaptation Programmes of Action (NAPA)—38 NAPAs had been completed as of October 2008. The NAPAs communicate the country's priority activities addressing the urgent and immediate needs and concerns relating to adaptation to the adverse effects of climate change.

In order to provide an in-depth example of a climate change adaptation effort outside of the United States, we selected the United Kingdom as a case study to better understand some of the actions that government officials can take to adapt to climate change. We selected the United Kingdom because it has initiated a coordinated climate change adaptation response at the national, regional, and local levels.

Over the past decade, the issuance of prominent reports and the fallout from major weather events created awareness among government officials of the need for the United Kingdom to adapt to inevitable changes to the nation's climate. For example, in 2002, the London Climate Change Partnership, a stakeholder-led group coordinated by the Greater London Authority, issued a report called *London's Warming*, which detailed the expected impacts of climate change and the key challenges to addressing it.[44] In addition, the 2006 *Stern Review* of the economics of climate change helped accelerate the national government's efforts to adapt.[45] These and other reports show that the United Kingdom could experience a variety of climate change effects in the future, including dry summers, wet winters, coastal erosion, and sea level rise.

In fact, the United Kingdom is already experiencing severe weather events. For example, in 2006, a dry period brought about water restrictions in London. The following year, large-scale flooding in the United Kingdom highlighted the need to respond to climate change and led to the *Pitt Review*, which examined resilience to flooding in the United Kingdom.[46] In addition, the nation's insurance sector, which currently offers comprehensive flood insurance coverage, has raised concerns about the growing flood risk and asked for government action.

In response to these concerns, the United Kingdom enacted national climate change legislation in 2008.[47] The law requires the British Secretary of State for Environment, Food and Rural Affairs to report periodically to Parliament with a risk assessment of the current and predicted impacts of climate change and to propose programs and policies for climate change adaptation. The law also authorizes the national government to require certain public authorities, such as water companies, to report on their assessment of the current and predicted impact of climate change in relation to the authority's functions as well as their proposals and policies for adapting to climate change. According to Department for Environment, Food and Rural Affairs (DEFRA) officials, the government department responsible for leading action on adaptation, an independent expert subcommittee of the Committee on Climate Change is to provide technical advice and oversee these efforts. The United Kingdom is also working with the European Union to incorporate climate change into its decisions and policies.

In the United Kingdom, different levels of government report working together to ensure that climate change considerations are incorporated into decision making. For example, the Government Office for London chairs the national government's Local and Regional Adaptation Partnership Board, which aims to facilitate climate change adaptation at local and regional levels by highlighting best practices and encouraging information sharing among local and regional officials. According to DEFRA officials, the primary role of the national government is to provide information, raise awareness, and encourage others to take action, not dictate how to adapt. In response to the United Kingdom's 2008 Climate Change Act, DEFRA officials said they are preparing a national risk assessment and conducting economic analyses to quantify the costs and benefits of adaptive actions. DEFRA officials said that these steps are to assist adaptation efforts undertaken by the national government, local government officials, and the private sector.

Adaptation activities are driven in part by the use of national performance measures, which affect local funding, and national government programs, according to DEFRA officials. The national government recently introduced a national adaptation indicator, which measures how well local governments are adapting to climate change risk. Performance measured by this and other indicators is the basis for national grants to local governments. Individual government agencies are also developing and implementing their own plans to address climate change effects. For example, the Environment Agency, which is responsible for environmental protection in England and Wales, as well as flood defense and water resource management, has initiatives in place to reduce water use to increase resilience to drought. It is also addressing flood risk, most notably with the Thames Barrier, a series of flood gates that protect London from North Sea storms (see figure 6).

> The Thames Barrier is a flood control system designed by the Greater London Council to respond to severe floods in 1953. The Thames Estuary 2100 plan, which was released for public comment on March 31, 2009, was undertaken to determine whether London's flood control infrastructure, including the Thames Barrier, can continue to protect London given the

projections for sea level rise and expected development. The Environment Agency, which operates the barrier, relied on models of sea level rise to determine that continuation of current operations with some marginal improve-ments, such as using the barrier's gates' ability to "over-rotate," combined with other measures throughout the floodplain, would be sufficient at this time. The plan includes a monitoring component and a schedule to take further action later this century.

Source: GAO.

Figure 6. The Thames Barrier, London, United Kingdom

The United Kingdom's climate change initiatives are built around locally relevant information generated centrally by two primary sources. The United Kingdom Climate Impacts Programme (UKCIP), a primarily publicly funded program housed in Oxford University, generates stakeholder-centered climate change decision-making tools and facilitates responses to climate change. UKCIP works with national, regional, and local users of climate data to increase awareness and encourage actions. For example, Hampshire County, in southern England, used climate scenarios generated by UKCIP to complete a test of the county's sensitivity to weather and other emergency scenarios. The

Met Office Hadley Centre, a government-funded climate research center, generates climate science information and develops models. According to a United Kingdom official, the Met Office Hadley Centre generated the bulk of the science for the UK Climate Projections 2009, while UKCIP, among others, provided user guidance and training to facilitate the use of these data.[48]

Regional and international partnerships have also played a significant role in providing guidance to further climate change adaptation efforts in the United Kingdom. For example, Government Office for London officials told us that the Three Regions Climate Change Group (which includes the East of England, South East of England, and London) has set up a group to promote retrofitting of existing homes. The group produced a report, which provided a checklist for developers, case studies, a good practices guide, and a breakdown of the costs involved.[49] On an international scale, Greater London Authority officials stated that they are working with cities such as Tokyo, Toronto, and New York City to share knowledge about climate change adaptation. In addition, a Hampshire County Council official told us about the county's participation in the European Spatial Planning—Adapting to Climate Events project, which provided policy guidance and decision-making tools to governments from several countries on incorporating adaptation into planning decisions.

FEDERAL, STATE, AND LOCAL OFFICIALS FACE NUMEROUS CHALLENGES WHEN CONSIDERING CLIMATE CHANGE ADAPTATION EFFORTS

The challenges faced by federal, state, and local officials in their efforts to adapt fell into three categories, based on our analysis of questionnaire results, site visits, and available studies. First, available attention and resources are focused on more immediate needs, making it difficult for adaptation efforts to compete for limited funds. Second, insufficient site-specific data, such as local projections of expected changes, makes it hard to predict the impacts of climate change, and thus hard for officials to justify the current costs of adaptation efforts for potentially less certain future benefits. Third, adaptation efforts are constrained by a lack of clear roles and responsibilities among federal, state, and local agencies.

Competing Priorities Make It Difficult to Use Limited Funds on Adaptation Efforts

Competing priorities limit the ability of officials to respond to the impacts of climate change, based on our analysis of Web-based questionnaire results, site visits, and available studies. We asked federal, state, and local officials to rate specific challenges related to awareness and priorities as part of our questionnaire. Table 2 presents the percentage of federal, state, and local respondents who rated these challenges as very or extremely challenging in our questionnaire. Appendix III includes a more detailed summary of federal, state, and local officials' responses to the questionnaire.

The highest rated challenge identified by respondents was an overall lack of funding for adaptation efforts. This problem is coupled with the competing priorities of more immediate concerns.

Lack of funding: The government officials who responded to our questionnaire identified the lack of funding for adaptation efforts as both the top challenge related to awareness and priorities and the top overall challenge explored in our questionnaire. Several respondents wrote that lack of funding limited their ability to identify and respond to the impacts of climate change, with one noting, for example, that "we have the tools, but we just need the funding and leadership to act." A state official similarly said that "we need a large and dedicated funding source for adaptation. It's going to take 5 to 10 years of funding to get a body of information that will help planning in the long run. We need to start doing that planning and research now." Several studies also suggested that it will be difficult, if not impossible, for any agency to approach the tasks associated with adaptation without permanent, dedicated funding. For example, a recent federal report on adaptation options for climate-sensitive ecosystems and resources stated that a lack of sufficient resources may pose a significant barrier to adaptation efforts.[50]

Officials also cited lack of funding as a challenge during our site visits. For example, King County officials said that they do not have resources budgeted directly for addressing climate change. Instead, the county tries to meet its adaptation goals by shifting staff and reprioritizing goals. The county officials said it was difficult to take action without dedicated funding because some adaptation options are perceived to be very expensive, and that if available funding cannot support the consideration of adaptation options then the old ways of doing business would remain the norm.

Table 2. Percentage of Challenges Related to Awareness and Priorities Rated as Very or Extremely Challenging

How challenging are each of the following for officials when considering climate change adaptation efforts?	Total responses[a]	Percentage who rated as very or extremely challenging[b]
Lack of funding for adaptation efforts	179	83.8
Non-adaptation activities are higher priorities	180	71.1
Lack of clear priorities for allocating resources for adaptation activities	181	70.2
Lack of public awareness or knowledge of adaptation	184	61.4
Lack of a specific mandate to address climate change adaptation	182	57.7
Lack of awareness or knowledge of adaptation among government officials	182	57.7
Lack of clarity about what activities are considered adaptation	181	55.2
Difficult to define adaptation goals and performance metrics	181	55.8
Lack of qualified staff to work on adaptation efforts	181	50.3

Source: GAO.

[a] The total column represents the number of officials who answered each question using numerical ratings, ranging from (1) not at all challenging through (5) extremely challenging, out of the 187 respondents that completed the questionnaire.

[b] The percentage column represents the number of officials rating each challenge as (4) very challenging or (5) extremely challenging divided by the total number of numerical ratings submitted by officials for (1) not at all challenging through (5) extremely challenging.

Competing priorities: Respondents' concerns over an overall lack of funding for adaptation efforts was further substantiated, and perhaps explained, by their ratings of challenges related to the priority of adaptation relative to other concerns. Specifically, about 71 percent (128 of 180) of the respondents rated the challenge "non-adaptation activities are higher priorities" as very or extremely challenging. The responses of federal, state, and local respondents differed for this challenge. Specifically, about 79 percent (37 of 47) of state officials and nearly 76 percent (44 of 58) of local officials who responded to the question rated "non-adaptation activities are higher priorities" as very or extremely challenging, compared with about 61 percent (44 of 72) of the responding federal officials.[51]

Several federal, state, and local officials noted in their narrative comments in our questionnaire how difficult it is to convince managers of the need to plan for long-term adaptation when they are responsible for more urgent concerns that have short-term decision-making time frames. One federal official explained that "it all comes down to resource prioritization. Election and budget cycles complicate long-term planning such as adaptation will require. Without clear top-down leadership setting this as a priority, projects with benefits beyond the budget cycle tend to get raided to pay current-year bills to deliver results in this political cycle." Several other officials who responded to our questionnaire expressed similar sentiments. A recent NRC report similarly concluded that, in some cases, decision makers do not prioritize adaptation because they do not recognize the link to climate change in the day-to-day decisions that they make.[52]

Our August 2007 report on climate change on federal lands shows how climate change impacts compete for the attention of decision makers with more immediate priorities.[53] This report found that resource management agencies did not, at that time, make climate change a priority, nor did their agencies' strategic plans specifically address climate change. Resource managers explained that they had a wide range of responsibilities and that without their management designating climate change as a priority, they focused first on near-term priorities.

Our questionnaire results and site visits demonstrate that public awareness can play an important role in the prioritization of adaptation efforts. About 61 percent (113 of 184) of the officials who responded to our questionnaire rated "lack of public awareness or knowledge of adaptation" as either very or extremely challenging. The need to adapt to climate change is a complicated issue to communicate with the public because the impacts vary by location and may occur well into the future. For example, officials in Maryland told us that, while the public may be aware that climate change will affect the polar ice cap, people do not realize that it will also affect Maryland. New York City officials said that it is easier to engage the public once climate change effects are translated into specific concerns, such as subway flooding. They said the term climate change adaptation can seem too abstract to the public.

Lack of Site-Specific Information Limits Adaptation Efforts

As summarized in table 3 and corroborated by our site visits and available studies, a lack of site-specific information—including information about the future benefits of adaptation activities—limits the ability of officials to respond to the impacts of climate change. See appendix III for a more detailed summary of federal, state, and local officials' responses to our Web-based questionnaire.

These challenges generally fit into two main categories: (1) the difficulty in justifying the current costs of adaptation with limited information about future benefits and (2) translating climate data—such as projected temperature and precipitation changes—into information that officials need to make decisions.

Justifying current costs with limited information about future benefits: Respondents rated "justifying the current costs of adaptation efforts for potentially less certain future benefits" as the greatest challenge related to information and as the second greatest of all the challenges we asked about. They rated the "size and complexity of future climate change impacts" as the second greatest challenge related to information.[54] These concerns are not new. In fact, a 1993 report on climate change adaptation by the Congressional Office of Technology Assessment posed the following question within its overall discussion of the issue: "why adopt a policy today to adapt to a climate change effect that may not occur, for which there is significant uncertainty about impacts, and for which benefits of the anticipatory measure may not be seen for decades?"[55] Several officials shared similar reactions in written responses to our questionnaire. For example, one local official asked, "How do we justify added expenses in a period of limited resources when the benefits are not clear?"

While the costs of policies to mitigate and adapt to climate change may be considerable, it is difficult to estimate the costs of inaction—costs which could be much greater, according to a recent NRC report.[56] This report cites the long time horizon associated with climate change, coupled with deep uncertainties associated with forecasts and projections, among other issues, as aspects of climate change that are challenging for decision making. Several officials who responded to our questionnaire noted similar concerns. For example, one federal official stated that decision makers needed to confront "the reality that the future will not echo the past and that we will forever be managing under future uncertainty."

Of particular importance in adaptation are planning decisions involving physical infrastructure projects, which require large capital investments and which, by virtue of their anticipated lifespan, will have to be resilient to changes in climate for many decades.[57] The long lead time and long life of large infrastructure investments require such decisions to be made well before climate change effects are discernable. For example, the United Kingdom Environment Agency's Thames 2100 Plan, which was released for consultation in April 2009, maps out necessary maintenance and operations needs for the Thames Barrier until 2070, at which point major changes will be required. Since constructing flood gates is a long-term process (the current barrier was finished 30 years after officials first identified a need for it), officials said they need the information now, even if the threat will not materialize until later.

Translating climate data into site-specific information: The process of providing useful information to officials making decisions about adaptation can be summarized in several steps.

First, data from global-scale models must be "downscaled" to provide climate information at a geographic scale relevant to decision makers. About 74 percent (133 of 179) of the officials who responded to our questionnaire rated "availability of climate information at relevant scale (i.e., downscaled regional and local information)" as very or extremely challenging. In addition, according to one federal respondent, "until we better understand what the impacts of climate change will be at spatial (and temporal) scales below what the General Circulation Models predict for the global scale, it will be difficult to identify specific adaptation strategies that respond to specific impacts."[58]

Table 3. Percentage of Challenges Related to Information Rated as Very or Extremely Challenging

How challenging are each of the following for officials when considering climate change adaptation efforts?	Total responses[a]	Percentage who rated as very or extremely challenging[b]
Justifying the current costs of adaptation efforts for potentially less certain future benefits	179	79.3
Size and complexity of *future* climate change impacts	180	76.7
Translating available climate information (e.g., projected temperature, precipitation) into impacts at the local level (e.g., increased stream flow)	182	74.7
Availability of climate information at relevant scale (i.e., downscaled regional and local information)	179	74.3
Understanding the costs and benefits of adaptation efforts	180	70
Lack of information about thresholds (i.e., limits beyond which recovery is impossible or difficult)	175	64.6
Making management and policy decisions with uncertainty about future effects of climate change	184	64.1
Lack of baseline monitoring data to enable evaluation of adaptation actions (i.e., inability to detect change)	181	62.4
Lack of certainty about the timing of climate change impacts	180	57.2
Accessibility and usability of available information on climate impacts and adaptation	182	53.3
Size and complexity of *current* climate change impacts	179	48.6

Source: GAO.

[a] The total column represents the number of officials who answered each question using numerical ratings, ranging from (1) not at all challenging through (5) extremely challenging, out of the 187 respondents that completed the questionnaire.

[b] The percentage column represents the number of officials rating each challenge as (4) very challenging or (5) extremely challenging divided by the total number of numerical ratings submitted by officials for (1) not at all challenging through (5) extremely challenging.

Our August 2007 report on climate change on federal lands demonstrated that resource managers did not have sufficient site-specific information to plan for and manage the effects of climate change on the federal resources they oversee.[59] In particular, the managers lacked computational models for local projections of expected changes. For example, at that time, officials at the Florida Keys National Marine Sanctuary said that they lacked adequate modeling and scientific information to enable managers to predict change on a small scale, such as that occurring within the sanctuary. Without such models, the managers' options were limited to reacting to already-observed effects.

Second, climate information must be translated into impacts at the local level, such as increased stream flow. About 75 percent (136 of 182) of the respondents rated "translating available climate information (e.g., projected temperature, precipitation) into impacts at the local level (e.g., increased stream flow)" as very or extremely challenging. Some respondents and officials interviewed during our site visits said that it is challenging to link predicted temperature and precipitation changes to specific impacts. For example, one federal respondent said that "we often lack fundamental information on how ecological systems/species respond to non-climate change related anthropogenic stresses, let alone how they will respond to climate change." Such predictions may not easily or directly match the information needs that could inform management decisions. For example, Maryland officials told us they do not have information linking climate model information, such as temperature and precipitation changes, to biological impacts, such as changes to tidal marshes. Similarly, King County officials said they are not sure how to translate climate change information into effects on salmon recovery efforts. Specifically, they said that there is incomplete information about how climate change may affect stream temperatures, stream flows, and other factors important to salmon recovery.

However, multiple respondents said that it was not necessary to have specific, detailed, downscaled modeling to manage for adaptation in the short term. For example, one federal respondent said that although modeling projections will get better over time, there will always be elements of uncertainty in how systems and species will react to climate change. Interestingly, federal, state, and local respondents perceived the challenges posed by site-specific information needs differently. About 85 percent (60 of 71) of the federal officials that responded to the question rated "translating available climate information into impacts at the local level" as very or extremely challenging, compared to around 75 percent (35 of 47) of the state officials and around 66 percent (40 of 59) of the local officials who responded.

Third, local impacts must be translated into costs and benefits, since this information is required for many decision-making processes. Almost 70 percent (126 of 180) of the respondents to our questionnaire rated "understanding the costs and benefits of adaptation efforts" as very or extremely challenging. As noted by one local government respondent, it is important to understand the costs and benefits of adaptation efforts so they can be evaluated relative to other priorities. In addition, a federal respondent said that tradeoffs between costs and benefits are an important component to making decisions under uncertainty.

Fourth, decision makers need baseline monitoring data to evaluate adaptation actions over time. Nearly 62 percent (113 of 181) of the respondents to our questionnaire rated the "lack of baseline monitoring data to enable evaluation of adaptation actions (i.e., inability to detect change)" as very or extremely challenging, one of the lower ratings for this category of challenges. As summarized by a recent NRC report, officials will need site-specific and relevant baselines of environmental, social, and economic information against which past and current decisions can be monitored, assessed, and changed.[60] Future decision-making success will be judged on how quickly and effectively numerous ongoing decisions can be adjusted to changing circumstances. For example, according to Maryland officials, the state lacks baseline data on certain key Chesapeake Bay species such as blue crab and striped bass, so it will be difficult to determine how climate change will affect them or if proposed adaptation measures were successful. Similarly, our August 2007 report on climate change on federal lands showed that resource managers generally lacked detailed inventories and monitoring systems to provide them with an adequate baseline understanding of the plant and animal species that existed on the resources they manage.[61] Without such information, it was difficult for managers to determine whether observed changes were within the normal range of variability.

Adaptation Efforts Are Constrained by a Lack of Clear Roles and Responsibilities

A lack of clear roles and responsibilities for addressing adaptation across all levels of government limits adaptation efforts, based on our analysis of federal, state, and local officials' responses to our Web-based questionnaire, site visits, and relevant studies. Table 4 presents respondents' views on how challenging different aspects of the structure and operation of the federal government are to adaptation efforts. See appendix III for a more detailed summary of federal, state, and local officials' responses to our Web-based questionnaire.

These challenges are summarized in two general categories: (1) lack of clear roles and responsibilities and (2) federal activities that constrain adaptation efforts.

Lack of clear roles and responsibilities: "A lack of clear roles and responsibilities for addressing adaptation across all levels of government (i.e., adaptation is everyone's problem but nobody's direct responsibility)" was identified by respondents as the greatest challenge related to the structure and operation of the federal government. Several respondents elaborated on their rating. For example, according to one state official, "there is a power struggle between agencies and levels of government rather than a lack of clear roles. Everyone wants to take the lead rather than working together in a collaborative and cohesive way." One local official said he "can't emphasize enough how the lack of coordination between agencies at the federal (and state) level severely complicates our abilities at the local level." Several respondents also noted that there is no element within the federal government charged with facilitating a collaborative response. Our questionnaire results show that local and state respondents consider the lack of clear roles and responsibilities to be a greater challenge than do federal respondents. Specifically, about 80 percent (48 of 60) of local officials and about 67 percent (31 of 46) of state officials who responded to the question rated the lack of clear roles and responsibilities as either very or extremely challenging, compared with about 61 percent (42 of 69) of the responding federal officials.

This lack of coordination and "institutional fragmentation" are serious challenges to adaptation efforts because clear roles are necessary for a large-scale response to climate change. As stated by one local government respondent, agencies "have numerous, overlapping jurisdictions and authorities, many of which have different (sometimes competing) mandates. If left to plan independently, they'll either do no adaptation planning or, if they do, likely come up with very different (and potentially conflicting) adaptation priorities." A recent NRC report comes to similar conclusions, noting that collaboration among agencies can be impeded by different enabling laws, opposing missions, or incompatible budgetary rules.[62] Such barriers—whether formalized or implicit—can lead to disconnects, conflicts, and turf battles rather than productive cooperation, according to this report.

Table 4. Percentage of Challenges Related to the Structure and Operation of the Federal Government Rated as Very or Extremely Challenging

How challenging are each of the following for officials when considering climate change adaptation efforts?	Total responses[a]	Percentage who rated as very or extremely challenging[b]
Lack of clear roles and responsibilities for addressing adaptation across all levels of government (i.e., adaptation is everyone's problem but nobody's direct responsibility)	178	69.7
The authority and capability to adapt is spread among many federal agencies (i.e., institutional fragmentation)	176	58
Lack of federal guidance or policies on how to make decisions related to adaptation	176	52.3
Existing federal policies, programs, or practices that hinder adaptation efforts	150	42.7
Federal statutory, regulatory, or other legal constraints on adaptation efforts	152	36.2

Source: GAO.

[a]The total column represents the number of officials who answered each question using numerical ratings, ranging from (1) not at all challenging through (5) extremely challenging, out of the 187 respondents that completed the questionnaire.

[b]The percentage column represents the number of officials rating each challenge as (4) very challenging or (5) extremely challenging divided by the total number of numerical ratings submitted by officials for (1) not at all challenging through (5) extremely challenging.

About 52 percent (92 of 176) of the respondents to our questionnaire rated the "lack of federal guidance or policies on how to make decisions related to adaptation" as very or extremely challenging. Their views echo our August 2007 report, which noted that federal resource managers were constrained by limited guidance about whether or how to address climate change and, therefore, were uncertain about what actions, if any, they should take.[63] In general, resource managers from all of the agencies we reviewed for that report said that they needed specific guidance to incorporate climate change into their management actions and planning efforts. For example, officials from several federal land and water resource management agencies said that guidance would help resolve differences in their agencies about how to interpret broad resource management authorities with respect to climate change and give them an imperative to take action.

A recent federal report on adaptation options for climate-sensitive ecosystems and resources reinforced these points.[64] It noted that, as resource managers become aware of climate change and the challenges it poses, a major limitation is lack of guidance on what steps to take, especially guidance that is commensurate with agency cultures and the practical experiences that managers have accumulated from years of dealing with other stresses, such as droughts and fires.

Our questionnaire results indicate that local government respondents consider the lack of federal guidance to be a greater challenge than state or federal respondents. Specifically, about 65 percent (39 of 60) of local officials who responded to the question rated the "lack of federal guidance or policies on how to make decisions related to adaptation" as either very or extremely challenging, compared to about 41 percent (19 of 46) of state officials and nearly 49 percent (33 of 67) of the federal officials that responded.

Federal activities that constrain adaptation efforts: Another challenge related to the structure and operation of the federal government is the existence of federal policies, programs, or practices that hinder adaptation efforts. While not the top challenge in the category, "existing federal policies, programs, or practices that hinder adaptation efforts"—which was rated as very or extremely challenging by about 43 percent (64 of 150) of the officials who responded to our questionnaire—is an important issue, as indicated by a wealth of related written comments submitted by respondents, comments from officials at our site visits, and a number of related studies.

Our work shows how, at least in some instances, federal programs may limit adaptation efforts. Our 2007 climate change-related report on FEMA's

National Flood Insurance Program and the U.S. Department of Agriculture's (USDA) Federal Crop Insurance Corporation, which insures crops against drought or other weather disasters, contrasted the experience of public and private insurers.[65] We found that many major private insurers were incorporating some near-term elements of climate change into their risk management practices. In addition, we found that some private insurers were approaching climate change at a strategic level by publishing reports outlining the potential industrywide impacts and strategies to proactively address the issue. In contrast, our report noted that the agencies responsible for the nation's key federal insurance programs had done little to develop the kind of information needed to understand their programs' long-term exposure to climate change for a variety of reasons. As a FEMA official explained in that report, the National Flood Insurance Program is designed to assess and insure against current—not future—risks. Unlike the private sector, neither this program nor the Federal Crop Insurance Corporation had analyzed the potential impacts of an increase in the frequency or severity of weather-related events on their operations. At our site visit, Maryland officials told us that FEMA's outdated delineation of floodplains, as well as its failure to consider changes in floodplain boundaries due to sea level rise, is allowing development in areas that are vulnerable to sea level rise in Maryland because local governments rely on its maps for planning purposes. Both FEMA and USDA have taken recent steps to address these concerns and have committed to study these issues further and report to Congress, with USDA estimating completion by December 31, 2009.[66]

Officials who responded to our questionnaire also identified several federal laws that hinder climate change efforts. A state official noted that many federal laws such as the Endangered Species Act, the Clean Water Act, and the Clean Air Act were passed before recognition of the effects of climate change. A federal official stated that federal environmental laws may need to be amended to provide greater authority for agencies to practice adaptive management.[67] The official noted that federal laws promoting development may also warrant re-examination to the extent they provide incentives that run counter to prudent land and resource planning in the climate change context.

One federal respondent stated that federal laws, regulations, and policies assume that long-term climate is stable and that species, ecosystems, and water resources can be managed to maintain the status quo or to restore them to prior conditions. This official observed that these objectives may no longer be achievable as climate change intensifies in the coming decades. A state official similarly noted that because of the effects of climate change, maintenance of the resource management status quo in any given area may no longer be possible. Part of the problem may lie in the inherent tension between the order of legal frameworks and the relative chaos of natural systems, which one legal commentator explained as follows: "Lawyers like rules. We like enforceable rules. We want our rules to be optimal, tidy, and timeless.... Collaborative ecosystem management, by contrast, is often messy, elaborate, cumbersome, ad hoc, and defiantly unconventional."[68] Several officials who responded to our questionnaire expressed similar concerns related to climate change adaptation. For example, one federal official stated that existing laws "were built for the status quo, but we now must re-engineer the entire legal framework to deal with the ongoing, perpetual, and rapid change. A systems view is essential in order to manage change optimally."

FEDERAL EFFORTS TO INCREASE AWARENESS, PROVIDE RELEVANT INFORMATION, AND DEFINE RESPONSIBILITIES COULD HELP GOVERNMENT OFFICIALS MAKE DECISIONS ABOUT ADAPTATION

Potential federal actions for addressing challenges to adaptation efforts fall into three areas, based on our analysis of questionnaire results, site visits, and available studies: (1) federal training and education initiatives that could increase awareness among government officials and the public about the impacts of climate change and available adaptation strategies; (2) actions to provide and interpret site-specific information that could help officials understand the impacts of climate change at a scale that would enable them to respond; and (3) steps Congress and federal agencies could take to encourage adaptation by setting priorities and reevaluating programs that hinder adaptation efforts.

Federal Training and Education Initiatives Would Assist Adaptation Efforts

Federal training and education initiatives would assist adaptation efforts, based on our analysis of our Web-based questionnaire, site visits, and relevant studies. Table 5 presents potential federal government actions related to awareness and priorities as rated by federal, state, and local officials who responded to our questionnaire. See appendix III for a more detailed summary of federal, state, and local officials' responses to our Web-based questionnaire.

We present these potential federal actions in three general categories: (1) training programs that could help government officials to develop more effective and better coordinated adaptation programs; (2) development of specific policy options for government officials; and (3) public education efforts to increase the public's understanding of climate change issues and the need to begin investing in preparatory measures.

Training for government officials: Training efforts could help officials collaborate and share insights for developing and implementing adaptation initiatives. Respondents rated the "development of regional or local educational workshops for relevant officials that are tailored to their responsibilities" as the most useful potential federal government action related to awareness and priorities. According to one federal official, "it is clear that training and communication may be the two biggest hurdles we face. We have the capabilities to adapt and to forecast scenarios of change and potential impacts of alternative adaptation options. We lack the will to exercise this capacity. The lack of that will is traceable to ignorance, sometimes willfully maintained." This respondent calls for "a massive educational process... designed and implemented all the way from the top-end strategic thinkers down to the ranks of tactical implementers of change and adaptation options." Training on how to make decisions with uncertainty would be particularly useful for frontline actors, such as city and county governments. For example, Maryland held an interactive summit on building "coast-smart communities," which brought together federal, state, and local officials involved with planning decisions in coastal areas. The summit employed role-playing to introduce participants to critical issues faced by coastal communities as a result of climate change. In addition, New York City DEP officials noted that their membership in the Water Utility Climate Alliance provided them with an important way to exchange information with water managers from across the nation.

Several respondents said that the federal government could play an important role in training officials at all levels of government. For example, one state official said that "because so many of us are only in the early stages of becoming aware of this issue, I think that a well organized training where many people would be learning the same thing and in the same way is important." However, a different state official questioned whether federal training would be effective for state and local officials, explaining that federal officials may not have enough knowledge about specific state and local challenges. The official thought that a better option may be to hold regional conferences with diverse groups of federal, state, and local officials so that those who are not up to speed can observe and learn from those who are. Interestingly, about 84 percent (38 of 45) of the state officials and nearly 75 percent (53 of 71) of the federal officials who responded to the question rated the "development of regional or local educational workshops for relevant officials that are tailored to their responsibilities" as very or extremely useful, compared to about 67 percent (42 of 63) of the local officials that responded.

Development of lists of policy options for government officials: The development of lists of "no regrets" actions—actions in which the benefits exceed the costs under all future climate scenarios—and other potential adaptation policy options could inform officials about efforts that make sense to pursue today and are "worth doing anyway." The Intergovernmental Panel on Climate Change defines a "no regrets" policy as one that would generate net social and economic benefits irrespective of whether or not anthropogenic climate change occurs. Such policies could include energy conservation and efficiency programs or the construction of green roofs in urban areas to absorb rainwater and moderate the effects of heat waves.

About 73 percent (133 of 181) of the officials who responded to our questionnaire rated the "development of lists of 'no regrets' actions (i.e., actions in which the benefits exceed the costs under all future climate scenarios)" as either very or extremely useful. The costs of no regrets strategies may be easier to defend, and proposing such strategies could be a way to initiate discussions of additional adaptation efforts. Likewise, about 71 percent (129 of 181) of respondents rated the "development of a list of potential climate change adaptation policy options" as either very or extremely useful.

However, several respondents questioned whether national lists of adaptation options would be useful, noting that adaptation is inherently local or regional in nature. For example, one federal official said that "it is unclear that it would be possible to develop a list of actions that truly is no regrets for all scenarios, all places, and all interested parties." This view suggests that adaptation options—"no regrets" or otherwise—may vary based on the climate impacts observed or projected for different geographic areas. As stated by one local official, "a national list would need to collect options from all regions across many sectors to be useful."

Table 5. Percentage of Potential Federal Government Actions Related to Awareness and Priorities Rated as Very or Extremely Useful

How useful, if at all, would each of the following federal government actions be for officials in efforts to adapt to a changing climate?	Total responses[a]	Percentage who rated as very or extremely useful[b]
Development of regional or local educational workshops for relevant officials that are tailored to their responsibilities	182	74.7
Development of lists of "no regrets" actions (i.e., actions in which the benefits exceed the costs under all future climate scenarios)	181	73.5
Development of a list of potential climate change adaptation policy options	181	71.3
Creation of a campaign to educate the public about climate change adaptation	184	70.1
Training of relevant officials on adaptation issues	182	69.8
Creation of a recurring stakeholder forum to explore the interaction of climate science and adaptation practice	184	64.7
Prioritization of potential climate change adaptation options	183	61.7

Source: GAO.

[a] The total column represents the number of officials who answered each question using numerical ratings, ranging from (1) not at all useful through (5) extremely useful, out of the 187 respondents that completed the questionnaire.

[b] The percentage column represents the number of officials rating each potential action as (4) very useful or (5) extremely useful divided by the total numerical ratings submitted by officials for (1) not at all useful through (5) extremely useful.

Regarding the prioritization of potential adaptation policy options, about 62 percent (113 of 183) of the respondents rated the "prioritization of potential climate change adaptation options" as very or extremely useful, the lowest-rated potential action related to awareness and priorities. Several respondents were adamant that prioritization should occur at the local level because of the variability of local impacts, and others said that federal agencies should assist such efforts, but not direct them. According to one state official respondent, federal efforts "should recognize and meet the needs of states and local governments. They should not... dictate policy." Interestingly, local officials who responded to our questionnaire rated prioritization of policy options as more useful than federal or state officials. Specifically, about 75 percent (47 of 63) of the local officials who responded to the question said that federal prioritization of potential climate change adaptation options would be very or extremely useful, compared to nearly 57 percent (40 of 70) and about 51 percent (24 of 47) of federal and state officials, respectively.

Public education: About 70 percent (129 of 184) of the respondents rated the "creation of a campaign to educate the public about climate change adaptation" as very or extremely useful. A variety of federal, state, and local programs are trying to fill this void, at least in areas of the country that are actively addressing adaptation issues. For example, the Chesapeake Bay National Estuarine Research Reserve (partially funded by NOAA) provides education and training on climate change to the public and local officials in Maryland. Maryland state officials recently provided local officials and the public in Somerset County information on the effects of sea level rise during a workshop. The workshop highlighted the need to incorporate information about sea level rise in the county's land use plans, given that it is expected to inundate a significant part of the county. In addition, the University of Washington's Climate Impacts Group (CIG)—a program funded under NOAA's Regional Integrated Sciences and Assessment program—has been interacting with the public about climate change issues, including adaptation, for over 10 years, according to officials we interviewed as part of our site visit to King County, Washington. Considerable local media coverage of environmental issues has also assisted with public awareness in King County.

Federal Actions to Provide and Interpret Site-Specific Information Would Help Officials Implement Adaptation Efforts

Federal actions to provide and interpret site-specific information would help address challenges associated with adaptation efforts, based on our analysis of our Web-based questionnaire, site visits, and relevant studies. Table 6 presents potential federal government actions related to information as rated by federal, state, and local officials who responded to our questionnaire. See appendix III for a more detailed summary of federal, state, and local officials' responses to our Web-based questionnaire.

We discuss these potential federal actions below in three general categories: (1) the development of regional, state, and local climate change impact and vulnerability assessments; (2) the development of processes and tools to access, interpret, and apply climate information; and (3) the creation of a federal service to consolidate and deliver climate information to decision makers to inform adaptation efforts.

Developing impact and vulnerability assessments: Respondents rated the "development of state and local climate change impact and vulnerability assessments" as the most useful action the federal government could take related to information. The development of regional assessments was also rated as similarly useful by respondents. Such assessments allow officials to build adaptation strategies based on the best available knowledge about regional or local changes and how those changes may affect natural and human systems. Nearly 94 percent (43 of 46) of the state officials and about 83 percent (52 of 63) of the local officials who responded to the question rated the development of state and local climate change impact and vulnerability assessments as either very or extremely useful, compared to about 69 percent (49 of 71) of federal officials.

Officials at all of the sites we visited reported relying on impact and vulnerability assessments to drive policy development and focus on the most urgent adaptation needs. For example, King County officials told us that regional climate modeling information provided by CIG was used to conduct a vulnerability assessment of wastewater treatment facilities in the county. In addition, Maryland officials said that the state's coastal adaptation initiative relied on localized impact and vulnerability information provided by the Maryland Commission on Climate Change's Scientific and Technical Working

Group, a stakeholder working group consisting of scientists and other relevant stakeholders.

Development of processes and tools to help officials use information: About 80 percent (148 of 185) of respondents rated the "development of processes and tools to help access, interpret, and apply available climate information" as very or extremely useful. Even with available regional and local climate data, officials will need tools to interpret what the data mean for decision making. For example, CIG told us of the strong need for Web-based decision-making tools to translate climate impacts into information relevant for decision makers. King County's Department of Natural Resources and Parks has developed a tool that uses data generated by CIG to help wastewater facilities model flooding due to sea level rise and storms. United Kingdom officials noted that the Climate Impacts Programme provides similar tools to assist decision makers in the United Kingdom.

The identification and sharing of best practices from other jurisdictions could also help meet the information needs of decision makers. Around 80 percent (126 of 157) of respondents rated the "identification and sharing of best practices" as very or extremely important. Best practices refer to the processes, practices, and systems identified in organizations that performed exceptionally well and are widely recognized as improving performance and efficiency in specific areas. Based on a range of our prior work, we have found that successfully identifying and applying best practices can reduce expenses and improve organizational efficiency. Several officials who responded to our questionnaire said that learning the best practices of others could be useful in efforts to develop adaptation programs.

Federal climate service: About 61 percent (107 of 176) of respondents rated the "creation of a federal service to consolidate and deliver climate information to decision makers to inform adaptation efforts" as very or extremely useful. According to two pending bills in Congress that would establish a National Climate Service within NOAA, its purpose would be to advance understanding of climate variability and change at the global, national, and regional levels and support the development of adaptation and response plans by federal agencies and state, local, and tribal governments.

Table 6. Percentage of Potential Federal Government Actions Related to Information Rated as Very or Extremely Useful

How useful, if at all, would each of the following federal government actions be for officials in efforts to adapt to a changing climate?	Total responses[a]	Percentage who rated as very or extremely useful[b]
Development of *state and local* climate change impact and vulnerability assessments	183	80.3
Identification and sharing of best practices	157[c]	80.3
Development of processes and tools to help officials access, interpret, and apply available climate information	185	80.0
Development of *regional* climate change impact and vulnerability assessments	182	77.5
Creation of a federal service to consolidate and deliver climate information to decision makers to inform adaptation efforts	176	60.8
Development of an interactive stakeholder forum for information sharing	184	56.5

Source: GAO.

[a] The total column represents the number of officials who answered each question using numerical ratings, ranging from (1) not at all useful through (5) extremely useful, out of the 187 respondents that completed the questionnaire.

[b] The percentage column represents the number of officials rating each potential action as (4) very useful or (5) extremely useful divided by the total numerical ratings submitted by officials for (1) not at all useful through (5) extremely useful.

[c] As previously noted, 187 respondents completed our questionnaire overall. While the number of responses for each individual question generally ranged from 183 to 186, only 159 respondents answered this question. See appendix III for more details.

Respondents offered a range of potential strengths and weaknesses for such a service. Several said that a National Climate Service would help consolidate information and provide a single-information resource for local officials, and others said that it would be an improvement over the current ad hoc system. A climate service would avoid duplication and establish an agreed set of climate information with uniform methodologies, benchmarks, and metrics for decision making, according to some officials. According to one federal official, consolidating scientific, modeling, and analytical expertise and capacity could increase efficiency. Some officials similarly noted that with such consolidation of information, individual agencies, states, and local governments would not have to spend money obtaining climate data for their adaptation efforts. Others said that it would be advantageous to work from one source of information instead of different sources of varying quality. Importantly, some officials said that a National Climate Service would demonstrate a federal commitment to adaptation and provide a credible voice and guidance to decision makers.

Other respondents, however, were less enthusiastic. Some voiced skepticism about whether it was feasible to consolidate climate information, and others said that such a system would be too rigid and may get bogged down in lengthy review processes. Furthermore, certain officials said building such capacity may not be the most effective place to focus federal efforts because the information needs of decision makers vary so much by jurisdiction. Several officials noted that climate change is an issue that requires a multidisciplinary response and a single federal service may not be able to supply all of the necessary expertise. For example, one federal official stated that the information needs of Bureau of Reclamation water managers are quite different from the needs of Bureau of Land Management rangeland managers, which are different from the needs of all other resource management agencies and programs. The official said that it seems highly unlikely that a single federal service could effectively identify and address the diverse needs of multiple agencies. Several respondents also said that having one preeminent source for climate change information and modeling could stifle contrary ideas and alternative viewpoints. Finally, several officials who responded to our questionnaire were concerned that a National Climate Service could divert attention and resources from current adaptation efforts by reinventing duplicative processes without making use of existing structures.

A recent NRC report recommends that the federal government's adaptation efforts should be undertaken through a new integrated interagency initiative with both service and research elements, but that such an initiative should not be centralized in a single agency.[69] Doing so, according to this report, would disrupt existing relationships between agencies and their constituencies and formalize a separation between the emerging science of climate response and fundamental research on climate and the associated biological, social, and economic phenomena. Furthermore, the report states that a National Climate Service located in a single agency and modeled on the weather service would by itself be less than fully effective for meeting the national needs for climate-related decision support. The NRC report also notes that such a climate service would not be user-driven and so would likely fall short in providing needed information, identifying and meeting critical needs for research for and on decision support, and adapting adequately to changing information needs.

Congress and Federal Agencies Could Encourage Adaptation Efforts by Clarifying Roles and Responsibilities

Federal actions to clarify the roles and responsibilities for government agencies could encourage adaptation efforts, based on our analysis of questionnaire results, site visits, and available studies. Table 7 presents potential federal actions related to the structure and operation of the federal government, as rated by the federal, state, and local officials who responded to our Web-based questionnaire. See appendix III for a more detailed summary of federal, state, and local officials' responses to our Web-based questionnaire.

As discussed below, these potential federal actions can be grouped into three areas: (1) new national adaptation initiatives, (2) review of programs that hinder adaptation efforts, and (3) guidance for how to incorporate adaptation into existing decision-making processes.

New national adaptation initiatives: Our questionnaire results identified the "development of a national adaptation fund to provide a consistent funding stream for adaptation activities" as the most useful federal action related to the structure and operation of the federal government. This result is not surprising, given that lack of funding was identified as the greatest challenge to adaptation efforts. One local official said that "funding for local governments is absolutely required. Local budgets are tight and require external stimulus for any hope of adaptation strategies to be implemented." Several state respondents noted that none of the other potential policy options are maximally useful unless there is also consistent funding available to implement them. Overall, about 98 percent (45 of 46) of state officials and nearly 88 percent (56 of 64) of the local officials who responded to the question rated the development of a national adaptation fund to provide a consistent funding stream for adaptation activities as very or extremely useful, compared to about 71 percent (47 of 66) of federal officials.

Table 7. Percentage of Potential Federal Government Actions Related to the Structure and Operation of the Federal Government Rated as Very or Extremely Useful

How useful, if at all, would each of the following federal government actions be for officials in efforts to adapt to a changing climate?	Total responses[a]	Percentage who rated as very or extremely useful[b]
Development of a national adaptation fund to provide a consistent funding stream for adaptation activities	179	84.4
Development of a national adaptation strategy that defines federal government priorities and responsibilities	181	71.3
Review of existing programs to identify and modify policies and practices that hinder adaptation efforts	180	67.8
Issuance of guidance, policies, or procedures on how to incorporate adaptation into existing policy and management processes	180	65.6
Development of a climate change extension service to help share and explain available information	181	59.1
Creation of a centralized government structure to coordinate adaptation funding	166	53.6

Source: GAO.

[a] The total column represents the number of officials who answered each question using numerical ratings, ranging from (1) not at all useful through (5) extremely useful, out of the 187 respondents that completed the questionnaire.

[b] The percentage column represents the number of officials rating each potential action as (4) very useful or (5) extremely useful divided by the total numerical ratings submitted by officials for (1) not at all useful through (5) extremely useful.

About 71 percent (129 of 181) of the officials who responded to our questionnaire rated the "development of a national adaptation strategy that defines federal government priorities and responsibilities" as very or extremely useful. As noted by a federal official who responded to our questionnaire, the cost of responding to a changing climate will be paid one way or another—either through ad hoc responses to emergencies or through a coordinated effort at the federal level guided by the best foresight and planning afforded by the current science. According to this official, a strategic approach may cost less than reactive policies in the long term and could be more effective. Officials we spoke with at our site visits and officials who responded to our questionnaire said that a coordinated federal response would also demonstrate a federal commitment to adaptation.

About 59 percent (107 of 181) of respondents rated the "development of a climate change extension service to help share and explain available information" as very or extremely useful. A climate change extension service could operate in the same way as USDA's Cooperative State Research, Education, and Extension Service, with land grant universities and networks of local or regional offices staffed by experts providing useful, practical, and research-based information to agricultural producers, among others.[70] Such a service could be responsible for educating private citizens, city planners, and others at the local level whose responsibilities are climate sensitive. For example, Maryland Forest Service officials noted that the Maryland Cooperative Extension Service provides training and information on the significance of climate change. Several respondents cautioned that whatever is done at the federal level should be consistently and adequately funded.

About 54 percent (89 of 166) of respondents rated as very or extremely useful the "creation of a centralized government structure to coordinate adaptation funding." While some cautioned that such a structure could limit the flexibility of existing federal, state, and local programs, others said that there was a need for more coordinated funding. Support for the idea, however, varied by level of government. Specifically, about 73 percent of the local (41 of 56) and almost 55 percent of the state (23 of 42) officials that responded to this question rated the "creation of a centralized federal government structure to coordinate adaptation funding" as either very or extremely useful, compared to only about 35 percent of the federal (23 of 65) respondents.

Reviewing programs that hinder adaptation: About 68 percent (122 of 180) of the respondents said it would be very or extremely useful to systematically review the kind of programs, policies, and practices discussed earlier in this report that may hinder adaptation efforts. Nearly 75 percent (46 of 61) of the local officials and about 70 percent (32 of 46) of the state officials who responded to the question rated the "review of existing programs to identify and modify policies and practices that hinder adaptation efforts" as very or extremely useful, compared to about 59 percent (41 of 70) of federal officials. One state official urged a review of both programs and laws, stating that "entrenched practices must be adapted to new realities." Our May 2008 report on the economics of climate change also identified actions that could assist officials in their efforts to adapt to climate change.[71] Some of the economists surveyed for that report suggested reforming insurance subsidy programs in areas vulnerable to natural disasters like hurricanes or flooding. Several noted that a clear federal role exists for certain sectors, such as water resource management, which could require additional resources for infrastructure development, research, and managing federal lands.

Federal, state, and local respondents also pointed to a number of federal laws as assisting adaptation efforts. For example, multiple officials cited the Global Change Research Act of 1990, which established a federal interagency research program to assist the United States and the world to understand, assess, predict, and respond to human-induced and natural processes of global change. Officials from the New York City Panel on Climate Change credited the 2001 Metro East Coast report issued for USGCRP with increasing awareness of regional climate change effects, which led to local government response.[72] Multiple officials also said that the National Environmental Policy Act could assist adaptation efforts by incorporating climate change adaptation into the assessment process. According to CEQ officials, the federal government could provide adaptation information under the National Environmental Policy Act provision that directs all federal agencies to make available to states, counties, municipalities, and others advice and information useful in restoring, maintaining, and enhancing the quality of the environment. According to certain officials, the Coastal Zone Management Act, which is administered by NOAA, could encourage adaptation to climate change at the state and local levels by allowing states and territories to develop specific coastal climate change plans or strategies. The state of Maryland is already using Coastal Zone Management Act programs to assess and respond to the risk of sea level rise and coastal hazards.

Guidance on how to consider adaptation in existing processes: Nearly 66 percent (118 of 180) of respondents rated the "issuance of guidance, policies, or procedures on how to incorporate adaptation into existing policy and management processes" as very or extremely useful. A federal respondent added that adapting to climate change means integrating adaptation strategies into the programs that are already ongoing and will rely upon the networks and institutions that already exist. These sentiments were echoed in a recent report, which suggested that the experience of deliberately incorporating climate adaptation into projects can be very helpful in developing a more systematic approach to adaptation planning and can serve as a kind of project-based policy development.[73] Furthermore, this report notes that leading programs integrate climate change adaptation into overarching policy documents such as official plans or policies. In the same vein, King County officials told us they work to "routinize" climate change into planning decisions and have incorporated climate change into the county's comprehensive plan. This plan, among other things, states that "King County should consider projected impacts of climate change, including more severe winter flooding, when updating disaster preparedness, levee investment, and land use plans, as well as development regulations."[74] Several respondents cautioned that federal guidance related to adaptation should be flexible enough to allow state and local governments to adapt their own approaches.

GOVERNMENTWIDE PLANNING AND COLLABORATION COULD ASSIST ADAPTATION EFFORTS

Climate change is a complex, interdisciplinary issue with the potential to affect every sector and level of government operations. Strategic planning is a way to respond to this governmentwide problem on a governmentwide scale. Our past work on crosscutting issues suggests that governmentwide strategic planning can integrate activities that span a wide array of federal, state, and local entities.[75] Strategic planning can also provide a comprehensive framework for considering organizational changes, making resource decisions, and holding officials accountable for achieving real and sustainable results.

As this report and others demonstrate, some communities and federal lands are already seeing the effects of climate change, and governments are beginning to respond.[76] However, as this report also illustrates, the federal government's emerging adaptation activities are carried out in an ad hoc manner and are not well coordinated across federal agencies, let alone state and local governments. Officials who responded to our questionnaire at all levels of government said that they face a range of challenges when considering adaptation efforts, including competing priorities, lack of site-specific data, and lack of clear roles and responsibilities. These officials also identified a number of potential federal actions that they thought could help address these challenges.

Multiple federal agencies, as well as state and local governments, will have to work together to address these challenges and implement new initiatives. Yet, our past work on collaboration among federal agencies suggests that they will face a range of barriers in doing so.[77] Agency missions may not be mutually reinforcing or may even conflict with each other, making consensus on strategies and priorities difficult.

Incompatible procedures, processes, data, and computer systems also hinder collaboration. The resulting patchwork of programs and actions can waste scarce funds and limit the overall effectiveness of the federal effort. In addition, many federal programs were designed decades ago to address earlier challenges, informed by the conditions, technologies, management models, and organizational structures of past eras.[78] Based on our prior work, key practices that can help agencies enhance and sustain their collaborative efforts include[79]

- defining and articulating a common outcome;
- agreeing on roles and responsibilities;
- establishing compatible policies, procedures, and other means to operate across agency boundaries;
- identifying and addressing needs by leveraging resources; and
- developing mechanisms to monitor, evaluate, and report on results.

As we have previously reported, perhaps the single most important element of successful management improvement initiatives is the demonstrated commitment of top leaders to change.[80] Top leadership involvement and clear lines of accountability are critical to overcoming natural resistance to change, marshalling needed resources, and building and maintaining the commitment to new ways of doing business.

CONCLUSIONS

A key question for decision makers in both Congress and the administration is whether to start adapting now or to wait until the effects of climate change are more obvious and widespread. Given the complexity and potential magnitude of climate change and the lead time needed to adapt, preparing for these impacts now may reduce the need for far more costly steps in the decades to come.

Adaptation, however, will require making policy and management decisions that cut across traditional sectors, issues, and jurisdictional boundaries. It will mean developing new approaches to match new realities. Old ways of doing business—such as making decisions based on the assumed continuation of past climate conditions—will not work in a world affected by climate change.

Certain state and local authorities on the "front lines" of early adaptation efforts understand this new reality and are beginning to take action. Our analysis of these efforts, responses to our questionnaire, and available studies revealed that federal, state, and local officials face numerous challenges when considering adaptation efforts. To be effective, federal efforts to address these challenges must be coordinated and directed toward a common goal.

RECOMMENDATIONS FOR EXECUTIVE ACTION

We recommend that the appropriate entities within the Executive Office of the President, such as the Council on Environmental Quality and the Office of Science and Technology Policy, in consultation with relevant federal agencies, state and local governments, and key congressional committees of jurisdiction, develop a national strategic plan that will guide the nation's efforts to adapt to a changing climate. The plan should, among other things, (1) define federal priorities related to adaptation; (2) clarify roles, responsibilities, and working relationships among federal, state, and local governments; (3) identify mechanisms to increase the capacity of federal, state, and local agencies to incorporate information about current and potential climate change impacts into government decision making; (4) address how resources will be made available to implement the plan; and (5) build on and integrate ongoing federal planning efforts related to adaptation.

APPENDIX I: SCOPE AND METHODOLOGY

Our review (1) determines what actions, if any, federal, state, local, and international authorities are taking to adapt to a changing climate; (2) identifies the challenges, if any, that federal, state, and local officials reported facing in their efforts to adapt; and (3) identifies actions that Congress and federal agencies could take to help address these challenges. We also provide information about our prior work on responding to similarly complex, interdisciplinary issues.

To determine the actions federal authorities are taking to adapt to climate change, we obtained summaries of current and planned adaptation-related efforts from a broad range of federal agencies. Full summaries from federal agencies are provided in a supplement to this report (see GAO-10-114SP). We obtained these summaries from the federal agencies with assistance from the U.S. Global Change Research Program (USGCRP), formerly the United States Climate Change Science Program. USGCRP coordinates and integrates federal research on changes in the global environment and their implications for society. USGCRP collected submissions from 12 of the 13 departments and agencies that participate in its program (see app. II for more details).[81]

We also obtained a summary of adaptation-related efforts from the Federal Emergency Management Agency, part of the U.S. Department of Homeland Security, as a follow up to prior GAO work on climate change and the Federal Emergency Management Agency's National Flood Insurance Program. Because the U.S. Department of Homeland Security is not part of USGCRP, we solicited its submission directly.

Because we wanted to include current federal activities that the agencies themselves consider to be related to adaptation, we did not modify the content of these summaries, except to remove references to specific individuals. We also did not independently confirm the information in the summaries. In addition, because the request for summaries was made to a select group of federal agencies, the activities compiled in this report should not be considered a comprehensive list of all recent and ongoing climate change adaptation efforts across the federal government.

In addition to gathering summaries, we also conducted an Internet search to identify other federal, state, or local organizations that are taking action to adapt to a changing climate. This search also helped to identify challenges agencies face in their efforts to adapt, as well as actions the federal government could take, which are relevant to our second and third objectives. We searched the Web sites of relevant organizations and agencies, such as the Intergovernmental Panel on Climate Change, the Pew Center on Global Climate Change, the Coastal States Organization, and federal agencies such as the Environmental Protection Agency and the National Oceanic and Atmospheric Administration. We also conducted Internet searches using relevant key words, such as "climate change" and "climate change adaptation." We reviewed publicly available English- language documents related to adaptation efforts in the United States and other countries that we identified through our search.

To address our three objectives, we also conducted 13 open-ended interviews with a select group of organizations and agencies that are engaged in climate change adaptation activities. We selected them based on their level of involvement in the issue of climate change adaptation, as determined by (1) previous GAO work; (2) scoping interviews (a "snowball" technique); and (3) our search of the background literature. We attempted to speak with organizations that are working on climate change adaptation, as well as those that represent sectors affected by it. We generally focused on organizations and sectors that are working on this issue on a national level (rather than just in one city or region) and that have also worked closely with state and local officials. The organizations included the National Association of Clean Water Agencies, the H. John Heinz III Center for Science, Economics, and the Environment, ICLEI— Local Governments for Sustainability, and the Nature Conservancy, among others. In addition, we spoke with two academics who had a long-standing involvement with climate change issues at the national and international levels to gather additional background information on the issue. Because we spoke with a select group of organizations and individuals, we cannot generalize our results to those we did not interview.

In addition to asking our interviewees about the actions they are taking to address adaptation, we also asked them to identify other relevant reports or studies we should include in our work and other agencies or organizations that are engaged in adaptation activities (part of our "snowball" technique). We also asked what actions they thought the federal government and Congress could take to help in their efforts.

To determine the actions federal, state, local, and international authorities are taking to adapt to a changing climate, we also visited four sites where government officials are taking actions to adapt. We chose these sites because they were frequently mentioned in the background literature and scoping interviews as examples of locations that are implementing climate change adaptation and which may offer particularly useful insights into the types of actions governments can take to plan for climate change impacts. These sites are neither comprehensive nor representative of all state and local climate change adaptation efforts. They include New York City; King County, Washington; the state of Maryland; and the United Kingdom, focusing on the London region. We included an international site visit to examine how other countries are starting to adapt, and we specifically selected the United Kingdom because its climate change adaptation efforts were mentioned frequently in the background literature and scoping interviews and because it had already begun to implement these efforts at the national, regional, and local levels. During our site visits, we gathered information through interviews with officials and stakeholders, observation of adaptation efforts, and reviewed related documents. We also followed up with officials after our visits to gather additional information.

To describe the challenges that federal, state, and local officials face in their efforts to adapt and the potential actions that Congress and federal agencies could take to help address these challenges, we administered a Web-based questionnaire to a nonprobability sample of 274 federal, state, and local officials who were identified by their organizations to be knowledgeable about adaptation. To identify relevant potential respondents, we worked with organizations that represent federal, state, and local officials. Specifically, we worked with organizations such as USGCRP (federal), National Association of Clean Air Agencies (state), and Conference of Mayors (local), among others, and asked them to identify officials who are knowledgeable about climate change adaptation. These officials were generally identified through their involvement in climate change working groups within these organizations, which indicated a level of interest and knowledge of the issue. The officials were then contacted by their organization to describe the purpose of our questionnaire and to ask if they would participate. The names and e-mail addresses of those who agreed were then provided to GAO. The federal, state, and local officials who responded represent a diverse array of disciplines, including planners, scientists, and public health professionals; however, their responses cannot be generalized to officials who did not complete our questionnaire.

To develop the questionnaire, information was compiled from background literature and interviews we conducted with relevant organizations and officials. Using this information, we developed lists of challenges and potential actions the federal government could take to address them.

Using closed-ended questions, respondents were asked to rate several challenges and actions on 5 point Likert scales (the closed-ended questions are reproduced in app. III). We also included open-ended questions to give respondents an opportunity to tell us about challenges and potential federal actions that we did not ask about. Lastly, we included additional open-ended questions to gather opinions on a small number of related topics.

Because this was not a sample survey, it has no sampling errors. However, the practical difficulties of conducting any questionnaire may introduce errors, commonly known as nonsampling errors. For example, difficulties in interpreting a particular question, sources of information available to respondents, or analyzing data can introduce unwanted variability in the results. We took steps to minimize such nonsampling errors.

For example, social science survey specialists designed the questionnaire in collaboration with GAO staff who had subject matter expertise. Then, we sent a draft of the questionnaire to several federal, state, and local organizations for comment. In addition, we pretested it with local, state, and federal officials to check that (1) the questions were clear and unambiguous, (2) terminology was used correctly, (3) the questionnaire did not place an undue burden on agency officials, and (4) the questionnaire was comprehensive and unbiased. Based on these steps, we made necessary corrections and edits before it was administered. When we analyzed the data, an independent analyst checked all computer programs. Since this was a Web-based instrument, respondents entered their answers directly into the electronic questionnaire, eliminating the need to key data into a database, minimizing error.

We developed and administered a Web-based questionnaire accessible through a secure server. When we completed the final questionnaire, including content and form, we sent an e-mail announcement of the questionnaire to our nonprobability sample of 274 federal, state, and local officials on May 13, 2009. They were notified that the questionnaire was available online and were given unique passwords and usernames on May 28, 2009. We sent follow-up e-mail messages on June 4, June 8, and June 12, 2009, to those who had not yet responded. Then we contacted the remaining nonrespondents by telephone to encourage them to complete the questionnaire online, starting on June 24, 2009. The questionnaire was available online until July 10, 2009. Questionnaires were completed by 187 officials, for a response rate of about 68 percent.[82] The response rate by level of government is about 82 percent for federal officials (72 out of 88), about 90 percent for state officials (47 out of 52), and about 50 percent (65 out of 131) for local officials.[83]

We presented our questionnaire results in six tables in our report, which show the relative rankings of the challenges and potential actions listed in our questionnaire based on the percentage of respondents that rated them very or extremely challenging (for challenges) or very or extremely useful (for potential actions). Both the challenges and potential actions are organized into groups related to the following: (1) awareness and priorities, (2) information, and (3) the structure and operation of the federal government. Tables showing more detailed summaries of federal, state, and local officials' responses to the questionnaire are included in appendix III.

We conducted this performance audit from September 2008 to October 2009 in accordance with generally accepted government auditing standards. Those standards require that we plan and perform the audit to obtain sufficient, appropriate evidence to provide a reasonable basis for our findings and conclusions based on our audit objectives. We believe that the evidence obtained provides a reasonable basis for our findings and conclusions based on our audit objectives.

APPENDIX II: INFORMATION ON SELECTED FEDERAL EFFORTS TO ADAPT TO A CHANGING CLIMATE

We obtained information from 13 selected federal departments and agencies on their current and planned climate change adaptation efforts. We present this information in a supplement to this report to provide a more complete picture of the activities that federal agencies consider to be related to climate change adaptation than has been available publicly (see GAO-10-114SP). We obtained this information directly from the agencies participating in the U.S. Global Change Research Program.[84]

Importantly, we did not modify the content of the agency submissions (except to remove references to named individuals) or assess its validity. In addition, because this information represents the efforts of a selected group of federal agencies, the agency activities compiled in the supplement should not be considered a comprehensive list of all recent and ongoing climate change adaptation efforts across the federal government. Any questions about the information presented in the supplement should be directed to the agencies themselves.

See the following list for the departments and agencies included in the supplement to this report:

U.S. Department of Agriculture

- Agricultural Marketing Service
- Agricultural Research Service
- Cooperative State Research, Education, and Extension Service
- Economic Research Service
- Farm Service Agency
- Forest Service
- Natural Resources Conservation Service

U.S. Department of Commerce

- National Oceanic and Atmospheric Administration

U.S. Department of Defense

- Office of the Secretary of Defense
- Army
- Navy
- Air Force
- Marine Corps
- U.S. Army Corps of Engineers

U.S. Department of Energy

U.S. Department of Health and Human Services

- Centers for Disease Control and Prevention
- National Institutes of Health

U.S. Department of Homeland Security

- Federal Emergency Management Agency

U.S. Department of the Interior

U.S. Department of State and U.S. Agency for International Development

U.S. Department of Transportation

- Office of Transportation Policy

U.S. Environmental Protection Agency
National Aeronautics and Space Administration
National Science Foundation

APPENDIX III: SUMMARY OF FEDERAL, STATE, AND LOCAL OFFICIALS' RESPONSES TO WEB-BASED QUESTIONNAIRE

Table 8. All Officials' Rating of Challenges Related to Awareness and Priorities

How challenging are each of the following for officials when considering climate change adaptation efforts?

	(1) Not at all	(2) Slightly	(3) Moderately	(4) Very	(5) Extremely	Not applicable	Don't know / no response	Total responses[a]	Average[b]
Lack of funding for adaptation efforts	0	4	25	43	107	1	3	183	4.41
Non-adaptation activities are higher priorities	4	15	33	62	66	5	1	186	3.95
Lack of clear priorities for allocating resources for adaptation activities	3	12	39	71	56	2	3	186	3.91
Lack of public awareness or knowledge of adaptation	0	20	51	83	30	0	2	186	3.67
Lack of awareness or knowledge of adaptation among government officials	2	17	58	74	31	0	2	184	3.63
Difficult to define adaptation goals and performance metrics	1	21	58	66	35	0	5	186	3.62

	(1) Not at all	(2) Slightly	(3) Moderate ly	(4) Very	(5) Extremel y	Not applicabl e	Don't know / no response	Total responses[a]	Average[b]
Lack of qualified staff to work on adaptation efforts	5	25	60	44	47	0	5	186	3.57
Lack of a specific mandate to address climate change adaptation	18	24	35	50	55	2	2	186	3.55
Lack of clarity about what activities are considered adaptation	3	19	59	79	21	2	2	185	3.53

Source: GAO.

[a] The total column represents the number of officials who answered each question out of the 187 respondents that completed the questionnaire.
[b] The average column represents the average of the numerical ratings submitted by officials for (1) not at all challenging through (5) extremely challenging.

Table 9. All Officials' Rating of Challenges Related to Information

How challenging are each of the following for officials when considering climate change adaptation efforts?

	(1) Not at all	(2) Slightly	(3) Moderately	(4) Very	(5) Extremely	Not applicable	Don't know/no response	Total responses[a]	Average[b]
Size and complexity of *future* climate change impacts	1	8	33	65	73	1	4	185	4.12
Justifying the current costs of adaptation efforts for potentially less certain future benefits	1	7	29	76	66	2	4	185	4.11
Translating available climate information (e.g., projected temperature, precipitation) into impacts at the local level (e.g., increased stream flow)	3	13	30	62	74	1	2	185	4.05
Availability of climate information at relevant scale (i.e., downscaled regional and local information)	4	15	27	66	67	0	4	183	3.99
Understanding the costs and benefits of adaptation efforts	0	5	49	78	48	2	3	185	3.94

Table 9 (Continued)

Making management and policy decisions with uncertainty about future effects of climate change	2	14	50	68	50	1	185	3.82
Lack of information about thresholds (i.e., limits beyond which recovery is impossible or difficult)	7	17	38	66	47	3	185	3.74
Lack of baseline monitoring data to enable evaluation of adaptation actions (i.e., inability to detect change)	7	17	44	78	35	1	184	3.65
Lack of certainty about the timing of climate change impacts	3	16	58	68	35	0	183	3.64
Accessibility and usability of available information on climate impacts and adaptation	6	25	54	64	33	0	184	3.51
Size and complexity of *current* climate change impacts	6	22	64	56	31	1	184	3.47

Source: GAO.

[a]The total column represents the number of officials who answered each question out of the 187 respondents that completed the questionnaire.

[b]The average column represents the average of the numerical ratings submitted by officials for (1) not at all challenging through (5) extremely challenging.

Table 10. All Officials' Rating of Challenges Related to the Structure and Operation of the Federal Government

How challenging are each of the following for officials when considering climate change adaptation efforts?	(1) Not at all	(2) Slightly	(3) Moderately	(4) Very	(5) Extremely	Not applicable	Don't know/ no response	Total responses[a]	Average[b]
Lack of clear roles and responsibilities for addressing adaptation across all levels of government (i.e., adaptation is everyone's problem but nobody's direct responsibility)	4	16	34	54	70	2	5	185	3.96
The authority and capability to adapt is spread among many federal agencies (i.e., institutional fragmentation)	4	23	47	66	36	2	7	185	3.61
Lack of federal guidance or policies on how to make decisions related to adaptation	11	22	51	53	39	3	6	185	3.49
Existing federal policies, programs, or practices that hinder adaptation efforts	8	31	47	30	34	3	31	184	3.34
Federal statutory, regulatory, or other legal constraints on adaptation efforts	14	33	50	29	26	4	29	185	3.13

Source: GAO.

[a] The total column represents the number of officials who answered each question out of the 187 respondents that completed the questionnaire.
[b] The average column represents the average of the numerical ratings submitted by officials for (1) not at all challenging through (5) extremely challenging.

Table 11. All Officials' Rating of Potential Federal Government Actions Related to Awareness and Priorities.

How useful, if at all, would each of the following federal government actions be for officials in efforts to adapt to a changing climate?

	(1) Not at all	(2) Slightly	(3) Moderately	(4) Very	(5) Extremely	Don't know/ no response	Total responses[a]	Average[b]
Development of regional or local educational workshops for relevant officials that are tailored to their responsibilities	3	7	36	64	72	2	184	4.07
Development of lists of "no regrets" actions (i.e., actions in which the benefits exceed the costs under all future climate scenarios)	4	13	31	60	73	5	186	4.02
Creation of a campaign to educate the public about climate change adaptation	1	19	35	60	69	0	184	3.96
Development of a list of potential climate change adaptation policy options	2	12	38	73	56	4	185	3.93
Training of relevant officials on adaptation issues	3	14	38	69	58	2	184	3.91
Creation of a recurring stakeholder forum to explore the interaction of climate science and adaptation practice	3	21	41	70	49	2	186	3.77
Prioritization of potential climate change adaptation options	9	19	42	70	43	3	186	3.65

Source: GAO.

[a] The total column represents the number of officials who answered each question out of the 187 respondents that completed the questionnaire.

[b] The average column represents the average of the numerical ratings submitted by officials for (1) not at all useful through (5) extremely useful.

Table 12. All Officials' Rating of Potential Federal Government Actions Related to Information

How useful, if at all, would each of the following federal government actions be for officials in efforts to adapt to a changing climate?	(1) Not at all	(2) Slightly	(3) Moderately	(4) Very	(5) Extremely	Don't know/ no response	Total responses[a]	Average[b]
Development of state and local climate change impact and vulnerability assessments	2	9	25	56	91	1	184	4.23
Development of regional climate change impact and vulnerability assessments	0	5	36	60	81	3	185	4.19
Development of processes and tools to help officials access, interpret, and apply available climate information	0	7	30	72	76	0	185	4.17
Identification and sharing of best practices	0	7	24	65	61	2	159[c]	4.15
Creation of a federal service to consolidate and deliver climate information to decision makers to inform adaptation efforts	11	20	38	41	66	9	185	3.74
Development of an interactive stakeholder forum for information sharing	1	23	56	58	46	1	185	3.68

Source: GAO.

[a]The total column represents the number of officials who answered each question out of the 187 respondents that completed the questionnaire.

[b]The average column represents the average of the numerical ratings submitted by officials for (1) not at all useful through (5) extremely useful.

[c]As previously noted, 187 respondents completed our questionnaire overall. While the number of responses for each individual question generally ranged from 183 to 186, only 159 respondents answered this question.

Table 13. All Officials' Rating of Potential Federal Government Actions Related to the Structure and Operation of the Federal Government

How useful, if at all, would each of the following federal government actions be for officials in efforts to adapt to a changing climate?	(1) Not at all	(2) Slightly	(3) Moderately	(4) Very	(5) Extremely	Don't know/ no response	Total responses[a]	Average[b]
Development of a national adaptation fund to provide a consistent funding stream for adaptation activities	7	8	13	38	113	5	184	4.35
Development of a national adaptation strategy that defines federal government priorities and responsibilities	4	12	36	65	64	4	185	3.96
Review of existing programs to identify and modify policies and practices that hinder adaptation efforts	1	19	38	65	57	5	185	3.88
Issuance of guidance, policies, or procedures on how to incorporate adaptation into existing policy and management processes	2	15	45	78	40	4	184	3.77
Development of a climate change extension service to help share and explain available information	8	20	46	54	53	3	184	3.69
Creation of a centralized government structure to coordinate adaptation funding	24	20	33	44	45	19	185	3.40

Source: GAO.

[a] The total column represents the number of officials who answered each question out of the 187 respondents that completed the questionnaire.
[b] The average column represents the average of the numerical ratings submitted by officials for (1) not at all useful through (5) extremely useful.

APPENDIX IV: COMMENTS FROM THE COUNCIL ON ENVIRONMENTAL QUALITY

EXECUTIVE OFFICE OF THE PRESIDENT
COUNCIL ON ENVIRONMENTAL QUALITY
WASHINGTON, D.C. 20503

John B. Stephenson
Director
Natural Resources and Environment
U.S. Government Accountability Office
441 G Street N.W.
Washington, DC 20548

Dear Mr. Stephenson,

Thank you for the opportunity to review and comment on Government Accountability Office's report, "Climate Change Adaptation: Strategic Federal Planning Could Help Government Officials Make More Informed Decisions." We circulated the report to the Climate Change Adaptation inter-agency committee for review and comment. The committee includes representatives from more than twelve agencies. We have also provided technical comments under separate cover.

We agree that adaptation is a critical area for federal government activity and think this report is a timely review of the subject. Overall, we agree with the main recommendation, that leadership and coordination is necessary within the federal government to ensure an effective and appropriate adaptation response. Further, we agree that this will help to catalyze the local, state and regional activities that are so critical to adaptation.

We have three main areas of concern with the report. First, we believe that the relative inexperience of the federal government on adaptation combined with the survey methodology used in this report may produce misleading results. Second, we believe that the report confuses the issue of cost/benefit analysis and scientific uncertainty. Third, we think the overall report does not focus enough on implementation challenges and recommendations.

Methodology
The report uses a survey methodology to assess relative roles and tasks for the federal government on adaptation. Survey respondents were selected from both within and outside the federal government, and all had experience with adaptation. However, the report also documents the relatively low level of activity within the federal government on adaptation, suggesting that most federal government respondents must be relatively inexperienced with adaptation issues. This is reinforced by the significant differences in some survey responses between respondents within the federal government, and those with presumably greater adaptation experience, outside of the federal government.

As a result, some of the survey findings appear to be questionable. For example, the survey found that developing a list of no-regrets actions would be a valuable product for the federal government to produce. While no-regrets actions are a critical part of adaptation, the variability and local nature of adaptation makes a federally produced list of no-regrets actions very difficult and possibly of limited utility. Therefore, while the

EXECUTIVE OFFICE OF THE PRESIDENT
COUNCIL ON ENVIRONMENTAL QUALITY
WASHINGTON, D.C. 20503

John B. Stephenson
Director
Natural Resources and Environment
U.S. Government Accountability Office
441 G Street N.W.
Washington, DC 20548

Dear Mr. Stephenson,

Thank you for the opportunity to review and comment on Government Accountability Office's report, "Climate Change Adaptation: Strategic Federal Planning Could Help Government Officials Make More Informed Decisions." We circulated the report to the Climate Change Adaptation inter-agency committee for review and comment. The committee includes representatives from more than twelve agencies. We have also provided technical comments under separate cover.

We agree that adaptation is a critical area for federal government activity and think this report is a timely review of the subject. Overall, we agree with the main recommendation, that leadership and coordination is necessary within the federal government to ensure an effective and appropriate adaptation response. Further, we agree that this will help to catalyze the local, state and regional activities that are so critical to adaptation.

We have three main areas of concern with the report. First, we believe that the relative inexperience of the federal government on adaptation combined with the survey methodology used in this report may produce misleading results. Second, we believe that the report confuses the issue of cost/benefit analysis and scientific uncertainty. Third, we think the overall report does not focus enough on implementation challenges and recommendations.

Methodology
The report uses a survey methodology to assess relative roles and tasks for the federal government on adaptation. Survey respondents were selected from both within and outside the federal government, and all had experience with adaptation. However, the report also documents the relatively low level of activity within the federal government on adaptation, suggesting that most federal government respondents must be relatively inexperienced with adaptation issues. This is reinforced by the significant differences in some survey responses between respondents within the federal government, and those with presumably greater adaptation experience, outside of the federal government.

As a result, some of the survey findings appear to be questionable. For example, the survey found that developing a list of no-regrets actions would be a valuable product for the federal government to produce. While no-regrets actions are a critical part of adaptation, the variability and local nature of adaptation makes a federally produced list of no-regrets actions very difficult and possibly of limited utility. Therefore, while the

EXECUTIVE OFFICE OF THE PRESIDENT
COUNCIL ON ENVIRONMENTAL QUALITY
WASHINGTON, D.C. 20503

survey results are an accurate reflection of respondents thinking, they do not necessarily paint the best roadmap for federal government action.

Cost/Benefit and Uncertainty
The report identifies "justifying current costs with limited information about future benefits" as a challenge to adaptation policy. The discussion of this challenge focuses on the scientific uncertainty inherent in climate projections as the main stumbling block for cost/benefit analysis. The section does not include other challenges identified in the survey, such as "understanding costs and benefits" of adaptive actions, or the challenge of prioritizing adaptation against other near-term actions. The survey written comments point out that given the scientific uncertainty on impacts, cost/benefit analysis is particularly important. In these cases, cost/benefit analysis is a separate concern to scientific uncertainty.

Our interpretation of these survey responses is that while scientific uncertainty is a concern and challenge for adaptation planning and implementation, there is also difficulty doing cost/benefit analysis. This difficulty could be addressed through providing decision-maker tools, like scenario analyses, and tools that help to quantify the cost and benefits of inaction and action.

Planning vs. Implementation
The recommendation focuses on 4 components of a national strategic adaptation plan: priorities, roles and responsibilities, information and planning. These are critical elements of a national strategy on adaptation, and respond to the main challenges identified in the report.

However, the report does not analyze the primary barriers or challenges to implementation, nor does make any recommendations on implementing adaptation. Experience to date on adaptation suggests that planning is critical, but that it does not guarantee implementation. Many of the challenges described in the survey could apply equally to implementation (e.g., public awareness) and some were specifically focused on implementation (e.g., funding). But implementation challenges are neither discussed nor developed in the report. Simply fulfilling the recommendations on planning will not be sufficient to help the US adapt to climate change.

Thank you for the opportunity to review this report prior to its publication.

Sincerely,

Maria Blair
Deputy Associate Director for Climate Change Adaptation

End Notes

[1] Major greenhouse gases include carbon dioxide (CO2); methane (CH4); nitrous oxide (N2O); and synthetic gases such as hydrofluorocarbons (HFC), perfluorocarbons (PFC), and sulfur hexafluoride (SF6).

[2] Statement of Dr. John P. Holdren, Director, Office of Science and Technology, Executive Office of the President before the Committee on Agriculture, United States Senate (Washington, D.C., July 22, 2009).

[3] See, e.g., National Climate Service Act of 2009, H.R. 2306, 111th Congress (2009); American Clean Energy and Security Act of 2009, H.R. 2454, 111th Congress (2009); National Climate Service Act of 2009, H.R. 2407, 111th Congress (2009).

[4] Information on selected federal efforts to adapt to climate change is provided in a supplement to this report (see GAO-10-114SP).

[5] Not all officials responded to every question.

[6] Secretarial Order No. 3289 (Sep. 14, 2009).

[7] The Executive Order required the U.S. Department of Agriculture, U.S. Department of Defense, EPA, Interior, and the U.S. Department of Commerce to submit draft reports. Draft reports are available at http://executiveorder.chesapeakebay.net/.

[8] *The Role of Federal Lands in Combating Climate Change: Hearing Before the Subcommittee on National Parks, Forests, and Public Lands of the House Committee on Natural Resources,* 111th Cong. 7-12 (2009) (written statement of Abigail Kimbell, Chief, U.S. Forest Service). Also, on January 16, 2009, the Forest Service issued guidance for addressing climate change considerations in land management planning and project implementation.

[9] For more information about Interior's Climate Change Task Force, see http://www.usgs.gov/global_change/doi taskforce.asp.

[10] The Climate Change Science Program is now referred to as the United States Global Change Research Program. For report citation, see S.H. Julius, J.M. West (eds.), J.S. Baron, B. Griffith, L.A. Joyce, P. Kareiva, B.D. Keller, M.A. Palmer, C.H. Peterson, and J.M. Scott, *Preliminary Review of Adaptation Options for Climate-Sensitive Ecosystems and Resources,* Final Report, Synthesis and Assessment Product 4.4 (SAP 4.4), a report for the U.S. Climate Change Science Program and the Subcommittee on Global Change Research, U.S. Environmental Protection Agency, Washington, D.C., 2008.

[11] M.L. Corn, R.W. Gorte, G. Siekaniec, M. Bryan, D. Cleaves, K. O'Halloran, *Global Climate Change and Federal Lands: Two Cases,* a presentation hosted by the Congressional Research Service, 2009.

[12] In technical comments to this report, Interior pointed out that there are significant links between federal land and natural resource management and infrastructure design and operation. According to Interior, proper management of lands and natural resources can help protect human infrastructure and can be an adaptation strategy for human infrastructure in and of itself. For example, restoring coastal wetlands can help protect human infrastructure against storm surges, rising sea level, and erosion.

[13] EPA developed this guide in conjunction with NOAA, Rhode Island Sea Grant, and the International City/County Management Association. See http://coastalsmartgrowth.noaa.gov/.

[14] M. J. Savonis, V.R. Burkett, and J.R. Potter (eds.), *Impacts of Climate Change and Variability on Transportation Systems and Infrastructure: Gulf Coast Study, Phase I,* Synthesis and Assessment Product 4.7 (SAP 4.4), a report for the U.S. Climate Change Science Program and the Subcommittee on Global Change Research, U.S. Department of Transportation, Washington, D.C., 2008.

[15] GAO, *Climate Change: Financial Risks to Federal and Private Insurers in Coming Decades Are Potentially Significant,* GAO-07-285 (Washington, D.C.: Mar. 16, 2007).

[16] J.L. Gamble (ed.), K.L. Ebi, F.G. Sussman, T.J. Wilbanks, *Analyses of the Effects of Global Change on Human Health and Welfare and Human Systems*, Synthesis and Assessment Product 4.6 (SAP 4.6), a report for the U.S. Climate Change Science Program and the Subcommittee on Global Change Research, U.S. Environmental Protection Agency, Washington, D.C., 2008.

[17] National Defense Authorization Act for Fiscal Year 2008, Pub. L. No. 110-181, § 951, 122 Stat. 290 (2008).

[18] USAID, *Adapting to Climate Variability and Change: A Guidance Manual for Development Planning* (August 2007) and *Adapting to Coastal Climate Change: A Guidebook for Development Planners* (May 2009).

[19] In technical comments to this report, Interior also cited other programs that can assist in international adaptation, including (1) the Famine Early Warning System, which uses remote sensing to monitor floods and droughts in Africa, the Americas, and Afghanistan (USGS); (2) wildland fire cooperation with Mexico, Canada, Australia, and New Zealand (Bureau of Land Management, National Park Service, FWS, Bureau of Indian Affairs); (3) integrated water resource management, dam operations and safety, irrigation, flood control, water conservation in arid ecosystems, and hydrologic monitoring in Africa, Asia, and the Middle East (Bureau of Reclamation, USGS); (4) 30 sister park relationships with 20 countries that facilitate technical exchange and joint monitoring of protected ecosystems; (5) ecosystem monitoring, conservation of migratory and shared species with Mexico and Canada (FWS, National Park Service, Bureau of Land Management, USGS); and (6) conservation grants for elephants, rhinoceros, tigers, great apes, marine turtles, neotropical migratory birds, and waterfowl habitat (FWS).

[20] National Research Council of the National Academies, Panel on Strategies and Methods for Climate-Related Decision Support, Committee on the Human Dimensions of Global Change, *Informing Decisions in a Changing Climate* (Washington, D.C., 2009).

[21] California Natural Resources Agency, 2009 *California Climate Adaptation Strategy, Discussion Draft*.

[22] See Terri L. Cruce, *Adaptation Planning: What U.S. States and Localities are Doing*, a special report prepared for the Pew Center on Global Climate Change, November 2007 (updated August 2009), available at http://www.pewclimate.org/working-papers/adaptation and The H. John Heinz III Center for Science, Economics, and the Environment, *A Survey of Climate Change Adaptation Planning* (Washington, D.C., 2007), available at http://www.heinzctr.org/publications/meeting_reports.shtml. In addition, see Susanne C. Moser, *Good Morning, America! The Explosive U.S. Awakening to the Need for Adaptation*, a special report prepared at the request of the NOAA Coastal Services Center and the California Energy Commission, May 2009, available at http://www.csc.noaa.gov/publications/need-for-adaptation.pdf.

[23] Local Law No. 17 (2008) of City of New York, § 2.

[24] New York City Department of Environmental Protection Climate Change Program, with contributions by Columbia University's Center for Climate Systems Research and HydroQual Environmental Engineers & Scientists, P.C., *Report 1: Assessment and Action Plan—A Report Based on the Ongoing Work of the DEP Climate Change Task Force* (New York City, N.Y., 2008).

[25] Columbia Earth Institute, *Climate Change and a Global City: the Potential Consequences of Climate Variability and Change Metro East Coast*, a special report prepared at the request of the U.S. Global Change Research Program, July 2001.

[26] The first of these documents has been released. See NPCC, *Climate Risk Information* (New York City, N.Y., 2009).

[27] Lia Ossiander and Kevin Rennert, "Impacts of Climate Change on Washington State: Summary of Plenary Sessions" (prepared for *The Future Ain't What it Used to Be: Planning for Climate Disruption* conference in 2005, sponsored by King County, Seattle, Wash., October 2005).

[28] King County Ordinance 15728 (Apr. 25, 2007). The district is funded by a countywide ad valorem property tax levy of 10 cents per $1,000 assessed value.

[29] King County, *2007 Climate Plan* (Seattle, Wash., 2007).

[30] *See* King County Exec. Order No. PUT 7-8 (Mar. 22, 2006) (Executive Order on Land Use Strategies for Global Warming Preparedness); King County Exec. Order No. PUT 7-7 (Mar. 22, 2006) (Executive Order on Environmental Management Strategies for Global Warming Preparedness); King County Exec. Order No. PUT 7-10-1 (Aug. 31, 2007) (Evaluation of Climate Change Impacts through the State Environmental Policy Act).

[31] King County, *King County Comprehensive Plan 2008* (October 2008).

[32] University of Washington Climate Impacts Group, *The Washington Climate Change Impacts Assessment: Evaluating Washington's Future in a Changing Climate* (Seattle, Wash., 2009).

[33] Maryland Commission on Climate Change Adaptation and Response Working Group, *Comprehensive Strategy for Reducing Maryland's Vulnerability to Climate Change Phase I: Sea Level Rise and Coastal Storms* (Annapolis, Md., 2008).

[34] Maryland Commission on Climate Change, *Climate Action Plan* (Annapolis, Md., 2008).

[35] Maryland Commission on Climate Change Adaptation and Response Working Group, *Comprehensive Strategy for Reducing Maryland's Vulnerability to Climate Change Phase I: Sea Level Rise and Coastal Storms*.

[36] 2008 Md. Laws 304, *codified at* Md. Envir. § 16-201.

[37] 2008 Md. Laws 119, *codified at* Md. Nat. Res. § 8-1807. Critical areas are determined by local jurisdictions and approved by the Critical Area Commission for the Chesapeake and Atlantic Coastal Bays, but the initial planning area included all waters and lands under the Chesapeake Bay and Atlantic Coastal Bays and their tributaries and all land and water areas within 1,000 feet beyond the landward boundaries of state or private wetlands and heads of tides.

[38] Wanda Diane Cole, Maryland Eastern Shore Resource Conservation & Development Council, *Sea Level Rise: Technical Guidance for Dorchester County,* a special report prepared at the request of the Maryland Department of Natural Resources, March 2008; URS and RCQuinn Consulting, Inc., *Somerset County Maryland Rising Sea Level Guidance,* a special report prepared at the request of Somerset County, Maryland, Annapolis, Md., 2008; and CSA International Inc., *Sea Level Rise Response Strategy Worcester County, Maryland,* a special report prepared at the request of Worcester County, Maryland Department of Comprehensive Planning, September 2008.

[39] See http://shorelines.dnr.state.md.us. *Maryland Shorelines Online* is a coastal hazards Web portal, centralizing information and data on shoreline and coastal hazards management in Maryland.

[40] See http://shorelines.dnr.state.md.us/sc_online.asp.

[41] Intergovernmental Panel on Climate Change, *Climate Change 2007: Impacts, Adaptation and Vulnerability, Contribution of Working Group II to the Fourth Assessment Report of the Intergovernmental Panel on Climate Change* (Cambridge, United Kingdom, 2007).

[42] Government of Canada, *From Impacts to Adaptation: Canada in a Changing Climate 2007* (Ottawa, Ontario, 2008).

[43] Australian Government Department of Climate Change, *Climate Change Adaptation Actions for Local Government* (Canberra, Australia, 2009).

[44] London Climate Change Partnership. *London's Warming: The Impacts of Climate Change on London* (London, United Kingdom, November 2002).

[45] Nicholas Stern, *Stern Review: The Economics of Climate Change* (October 2006).

[46] Michael Pitt, *Pitt Review: Learning Lessons from the 2007 Floods* (June 2008).

[47] Climate Change Act 2008, ch. 27 (Eng.).

[48] The UK Climate Projections (UKCP09) provide climate information for the United Kingdom up to the end of this century. See http://ukcp09.defra.gov.uk/.

[49] *Your Home in a Changing Climate: Retrofitting Existing Homes for Climate Change Impacts*, a special report prepared at the request of the Three Regions Climate Change Group, February 2008.

[50] S.H. Julius, J.M. West (eds.), J.S. Baron, B. Griffith, L.A. Joyce, P. Kareiva, B.D. Keller, M.A. Palmer, C.H. Peterson, and J.M. Scott, *Preliminary Review of Adaptation Options for Climate-Sensitive Ecosystems and Resources*, Final Report, SAP 4.4.

[51] Differences by level of government (federal, state, and local) that are reported are for illustrative purposes and may not be statistically different. We present selected examples where the difference between federal, state, or local responses is greater than 15 percent and the difference presents useful context for the overall results. There were other differences by level of government that are not presented in this report.

[52] National Research Council of the National Academies, Panel on Strategies and Methods for Climate-Related Decision Support, Committee on the Human Dimensions of Global Change, *Informing Decisions in a Changing Climate*.

[53] GAO, *Climate Change: Agencies Should Develop Guidance for Addressing the Effects on Federal Land and Water Resources*, GAO-07-863 (Washington, D.C.: Aug. 7, 2007).

[54] About 77 percent of the officials who responded to our questionnaire rated the "size and complexity of *future* climate change impacts" as very or extremely challenging, whereas only about 49 percent of the officials rated the "size and complexity of *current* climate change impacts" similarly.

[55] While noting that it may be appealing to delay adaptation actions given uncertainty associated with where, when, and how much change will occur, the report also states that delay may leave the nation poorly prepared to deal with the changes that do occur and may increase the possibility of impacts that are irreversible or otherwise very costly. See U.S. Congress, Office of Technology Assessment, *Preparing for an Uncertain Climate—Volume I*, OTA-O-567 (Washington, D.C.: U.S. Government Printing Office, October 1993).

[56] National Research Council of the National Academies, Panel on Strategies and Methods for Climate-Related Decision Support, Committee on the Human Dimensions of Global Change, *Informing Decisions in a Changing Climate*.

[57] Government of Canada, From Impacts to Adaptation: Canada in a Changing Climate 2007 (Ottawa, Ontario, 2008).

[58] A General Circulation Model (GCM) is a global, three-dimensional computer model of the climate system which can be used to simulate human-induced climate change. GCMs are highly complex and they represent the effects of such factors as reflective and absorptive properties of atmospheric water vapor, greenhouse gas concentrations, clouds, annual and daily solar heating, ocean temperatures, and ice boundaries. The most recent GCMs include global representations of the atmosphere, oceans, and land surface.

[59] GAO-07-863.

[60] National Research Council of the National Academies, Panel on Strategies and Methods for Climate-Related Decision Support, Committee on the Human Dimensions of Global Change, *Informing Decisions in a Changing Climate*.

[61] GAO-07-863.

[62] National Research Council of the National Academies, Panel on Strategies and Methods for Climate-Related Decision Support, Committee on the Human Dimensions of Global Change, *Informing Decisions in a Changing Climate*.

[63] GAO-07-863.

[64] Julius, S.H., J.M. West (eds.), J.S. Baron, B. Griffith, L.A. Joyce, P. Kareiva, B.D. Keller, M.A. Palmer, C.H. Peterson, and J.M. Scott, *Preliminary Review of Adaptation Options for Climate-Sensitive Ecosystems and Resources*, Final Report, SAP 4.4.

[65] GAO-07-285.

[66] As mentioned, FEMA is currently conducting a study on the impact of climate change on the National Flood Insurance Program, which will be completed in March 2010. According to FEMA, this study will provide policy options and recommendations regarding the effects of

climate change on the National Flood Insurance Program. At USDA, the Risk Management Agency has contracted with a research group to provide a technical report on climate change impacts on the Federal Crop Insurance Corporation and develop a program impact model. The contractor has submitted preliminary results and the final report is due by December of this year. Using information contained in the report and other information, the Risk Management Agency will evaluate how it can adapt the crop insurance program to accommodate potential climate change scenarios.

[67] In 2004, NRC defined adaptive management as a process that promotes flexible decision making in the face of uncertainties, as outcomes from management actions and other events become better understood. See GAO, *Yellowstone Bison: Interagency Plan and Agencies' Management Need Improvement to Better Address Bison-Cattle Brucellosis Controversy*, GAO-08-291 (Washington, D.C.: Mar. 7, 2008). Adaptive management can be used to reduce the adverse effects of climate change on ecosystems. See C. Parmesan and H. Galbraith, *Observed Impacts of Global Climate Change in the U.S.* (2004). However, significant challenges confront those wishing to apply the technique to complex problems, such as addressing the effects of climate change on land use designations in land management plans prepared under the National Forest Management Act or the Federal Land Policy and Management Act of 1976. See R. Gregory et. al., "Deconstructing Adaptive Management: Criteria for Applications to Environmental Management," *Ecological Applications*, vol. 16, no. 6 (December 2006). Indeed, adaptive management "may be most difficult to implement in precisely those circumstances in which it is most needed." *Id.*

[68] Karkkainen, "Collaborative Ecosystem Governance: Scale, Complexity, and Dynamism," 21 *Va. Envtl. L.J.* 189, (2008): 243-35. Karkkainen's advice to lawyers who are unsettled by this apparent conflict is "let's get over it." *Id.* at 235.

[69] National Research Council of the National Academies, Panel on Strategies and Methods for Climate-Related Decision Support, Committee on the Human Dimensions of Global Change, *Informing Decisions in a Changing Climate*.

[70] See http://www.csrees.usda.gov/Extension/ for more information about USDA's extension service.

[71] GAO, Climate Change: Expert Opinion on the Economics of Policy Options to Address Climate Change, GAO-08-605 (Washington, D.C.: May 9, 2008).

[72] Columbia Earth Institute, *Climate Change and a Global City: The Potential Consequences of Climate Variability and Change Metro East Coast*, a special report prepared at the request of the U.S. Global Change Research Program, July 2001.

[73] The Clean Air Partnership, *Cities Preparing for Climate Change: A Study of Six Urban Regions* (May 2007).

[74] King County, *King County Comprehensive Plan 2008*.

[75] GAO, *A Call For Stewardship: Enhancing the Federal Government's Ability to Address Key Fiscal and Other 21st Century Challenges*, GAO-08-93SP (Washington, D.C.: Dec. 17, 2007).

[76] GAO, *Alaska Native Villages: Limited Progress Has Been Made on Relocating Villages Threatened by Flooding and Erosion*, GAO-09-551 (Washington, D.C.: June 3, 2009), and GAO-07-863.

[77] GAO, *Results-Oriented Government: Practices That Can Help Enhance and Sustain Collaboration among Federal Agencies*, GAO-06-15 (Washington, D.C.: Oct. 21, 2005), and *Managing for Results: Barriers to Interagency Coordination*, GAO/GGD-00-106 (Washington, D.C.: Mar. 29, 2000).

[78] GAO, *21st Century Challenges: Reexamining the Base of the Federal Government*, GAO-05-325SP (Washington, D.C.: Feb. 1, 2005).

[79] GAO-06-15.

[80] GAO, *Management Reform: Elements of Successful Improvement Initiatives*, GAO/T-GGD-00-26 (Washington, D.C.: Oct. 15, 1999).

[81] We did not receive a submission from the Smithsonian Institution.

[82] Not all officials responded to every question.
[83] Three officials from levels of government other than federal, state, or local—such as a regional level—also responded to the questionnaire.
[84] The U.S. Global Change Research Program (USGCRP) coordinates and integrates federal research on changes in the global environment and their implications for society. We did not receive a submission from the Smithsonian Institution. In addition to the agencies that participate in USGCRP, we also obtained a summary of current and planned adaptation-related efforts from the Federal Emergency Management Agency, part of the U.S. Department of Homeland Security, because of prior GAO adaptation-related work on its National Flood Insurance Program.

In: Climate Change Adaptation
Editor: Elizabeth N. Brewster

ISBN: 978-1-61728-889-0
© 2010 Nova Science Publishers, Inc.

Chapter 2

OPENING STATEMENT OF EDWARD J. MARKEY, BEFORE THE SELECT COMMITTEE ON ENERGY INDEPENDENCE AND GLOBAL WARMING, HEARING ON "BUILDING U.S. RESILIENCE TO GLOBAL WARMING IMPACTS"

We all remember the tragic consequences of Hurricane Katrina – the breached levees, water-filled streets, and families seeking shelter in the Superdome. While many individuals courageously responded to this disaster, government leadership failed the people of New Orleans when they needed help most. Katrina foreshadows the consequences of climate change if we do not make the necessary preparations.

Since then, scientists have shown that the warming of our climate system from emissions of heat-trapping gases – from our tailpipes and smokestacks – is unequivocal.

We face not only an increasing number of strong storms, but also many permanent alterations that will affect people throughout the country. Coastal cities like Boston will be at risk of inundation from sea level rise, which is accelerating as our oceans warm and our polar ice caps melt. Alaskan villages are finding the land they call home literally melting out from underneath them as the permafrost thaws. In the West, our shrinking mountain snowpack strains our water resource systems. Throughout this country, our farms are threatened by rising temperatures, water scarcity, and pests. For a projected 2.2 degree

(Fahrenheit) rise in temperatures over the next 30 years, we can expect significant declines in the crops that make up the base of our food system.

The past is no longer a predictor of the future. We need to develop our resilience in order to safeguard our health, our environment, our economy, and our national security. We need to develop a comprehensive strategy to adapt, conduct world-class climate research, and coordinate federal, state, and local action.

Now, some will argue that we should not address the root of the problem and only address its symptoms – that we should only adapt to climate change and not address global warming pollution. We cannot just address the symptoms. When someone has a heart attack, the doctor prescribes medication to help prevent another attack and puts the patient on a low-fat diet to improve long-term health. Our country experienced a heart attack in New Orleans and we must now develop BOTH the institutional medication to manage the impacts of warming AND ALSO shift society to a low-carbon energy regimen for a healthy climate. Just as we cannot medicate our way out of heart problems, we cannot simply adapt our way out of global warming.

We have taken the first steps to cut carbon pollution and build resilience to global warming impacts. Earlier this year, the House passed the Waxman-Markey American Clean Energy and Security Act, which will set us on a pollution cutting path and at the same time create millions of new jobs, making America the global leader of the clean energy economy. The Act will also create a National Climate Service that will provide decision-makers with the very best climate information and help federal agencies and states adapt to the dangerous consequences of climate change.

In a new report that I requested, the Government Accountability Office assesses the current steps our country is taking to address the impacts of global warming. They find that federal efforts thus far have been largely ad hoc. To effectively address the impacts, we need a strategic plan that sets our priorities, improves the information available to decision-makers, and clarifies the roles and responsibilities of federal, state, and local governments.

I look forward to the testimony of our witnesses and hearing from them how Congress can help build our resilience to global warming.

In: Climate Change Adaptation
Editor: Elizabeth N. Brewster

ISBN: 978-1-61728-889-0
© 2010 Nova Science Publishers, Inc.

Chapter 3

TESTIMONY OF JOHN B. STEPHENSON, DIRECTOR, NATURAL RESOURCES AND ENVIRONMENT, BEFORE THE SELECT COMMITTEE ON ENERGY INDEPENDENCE AND GLOBAL WARMING

October 22, 2009

Mr. Chairman and Members of the Committee:

I am pleased to be here today to discuss our report to this committee on climate change adaptation and the role strategic federal planning could play in government decision making. Changes in the climate attributable to increased concentrations of greenhouse gases may have significant impacts in the United States and internationally.[1] For example, climate change could threaten coastal areas with rising sea levels. In recent years, climate change adaptation—adjustments to natural or human systems in response to actual or expected climate change—has begun to receive more attention because the greenhouse gases already in the atmosphere are expected to continue altering the climate system into the future, regardless of efforts to control emissions. According to a recent report by the National Research Council (NRC), however, individuals and institutions whose futures will be affected by climate change are unprepared both conceptually and practically for meeting the challenges and opportunities it presents. In this context, adapting to climate change requires making policy and management decisions that cut across traditional economic sectors, jurisdictional boundaries, and levels of government. My testimony is based on our October 2009 report,[2] which is being publicly released today, and addresses three issues: (1) what actions federal, state, local, and international authorities are taking to adapt to a changing climate; (2) the challenges that federal, state, and local officials face in their efforts to adapt; and (3) the actions that Congress and federal agencies could take to help address these challenges. We also provide information about our prior work on similarly complex, interdisciplinary issues.

We employed a variety of methods to assess these issues. To determine the actions federal, state, local, and international authorities are taking to adapt to a changing climate, we obtained summaries of adaptation-related efforts from a broad range of federal agencies and visited four sites where government officials are taking actions to adapt. The four sites were New York City; King County, Washington; the state of Maryland; and the United Kingdom, focusing on London and Hampshire County. We gathered information during and after site visits through observation of adaptation efforts, interviews with officials and stakeholders, and a review of documents provided by these officials. To describe challenges that federal, state, and local officials face in their efforts to adapt and the actions that Congress and federal agencies could take to help address these challenges, we developed a Web-based questionnaire, and sent it to 274 federal, state, and local officials knowledgeable about adaptation.[3] Within the questionnaire, we organized questions about challenges and actions into groups related to the following: (1) awareness among governmental officials and the public about climate change impacts and setting priorities with respect to available adaptation strategies; (2) sufficiency of information to help officials understand climate change impacts at a scale that enables them to respond; and (3) the structure and operation of the federal government including whether roles and responsibilities were clear across different levels of government.

We conducted our review from September 2008 to October 2009 in accordance with generally accepted government auditing standards. A more detailed description of our scope and methodology is available in appendix I of our report.

Mr. Chairman, the following summarizes the findings on each of the issues discussed in our report:

- *Federal, state, local, and international efforts to adapt to climate change*: Although there is no coordinated national approach to adaptation, several federal agencies report that they have begun to take action with current and planned adaptation activities. These activities are largely ad hoc and fall into categories such as information for decision making, federal land and natural resource management, and governmentwide adaptation strategies, among others. For example, the National Oceanic and Atmospheric Administration's (NOAA) Regional Integrated Sciences and Assessments program supports climate change research to meet the needs of decision makers and policy planners at the national, regional, and local levels. In addition, several federal agencies have reported beginning to consider measures that would strengthen the resilience of natural resources in the face of climate change. For example, on September 14, 2009, the Department of the Interior issued an order designed to address the impacts of climate change on the nation's water, land, and other natural and cultural resources.[4] While no single entity is coordinating climate change adaptation efforts across the federal government, several federal entities are beginning to develop governmentwide strategies to adapt to climate change. For example, the President's Council on Environmental Quality (CEQ) is leading a new initiative to coordinate the federal response to climate change in conjunction with the Office of Science and Technology Policy, NOAA, and other agencies. Similarly, the U.S. Global Change Research Program, which coordinates and integrates federal research on climate change, has developed a series of "building blocks" that outline options for future climate change work, including science to inform adaptation.

- While many government authorities have not yet begun to adapt to climate change, some at the state and local levels are beginning to plan for and respond to climate change impacts. We visited three U. S. sites—New York City; King County, Washington; and the state of Maryland—where government officials are taking such steps. Our analysis of these sites suggests three major factors have led these governments to act. First, natural disasters such as floods, heat waves, droughts, or hurricanes raised public awareness of the costs of potential climate change impacts. Second, leaders in all three sites used legislation, executive orders, local ordinances, or action plans to focus attention and resources on climate change adaptation. Finally, each of the governments had access to relevant site-specific information to provide a basis for planning and management efforts. This site-specific information arose from partnerships that decision makers at all three sites formed with local universities and other government and nongovernment entities. Limited adaptation efforts are also taking root in other countries around the world. As in the case of the state and local efforts we describe, some of these adaptation efforts have been triggered by the recognition that current weather extremes and seasonal changes will become more frequent in the future. Our review of climate change adaptation efforts in the United Kingdom describes how different levels of government work together to ensure that climate change considerations are incorporated into decision making.

- *Government officials face numerous challenges when considering adaptation efforts:* The challenges faced by federal, state, and local officials in their efforts to adapt fall into the following three categories, based on our analysis of questionnaire results, site visits, and available studies:
- First, available attention and resources are focused on more immediate needs, making it difficult for adaptation efforts to compete for limited funds. For example, about 71 percent (128 of 180) of the officials who responded to our questionnaire rated "non-adaptation activities are higher priorities" as very or extremely challenging when considering climate change adaptation efforts.
- Second, insufficient site-specific data, such as local projections of expected changes, make it hard to predict the impacts of climate change, and thus hard for officials to justify the current costs of adaptation efforts for potentially less certain future benefits. For example, King County officials said they are not sure how to translate climate change information into effects on salmon recovery efforts.
- Third, adaptation efforts are constrained by a lack of clear roles and responsibilities among federal, state, and local agencies. Of particular note, about 70 percent (124 of 178) of the respondents rated the "lack of clear roles and responsibilities for addressing adaptation across all levels of government" as very or extremely challenging. Interestingly, local and state respondents rate this as a greater challenge than did federal respondents. About 80 percent (48 of 60) of local officials and about 67 percent (31 of 46) of state officials who responded to the question rated the issue as either very or extremely challenging, compared with about 61 percent (42 of 69) of the responding federal officials.[5]
- *Federal efforts could help government officials make decisions about adaptation:* Potential federal actions for addressing challenges to

adaptation efforts fall into the following three areas, based on our analysis of questionnaire results, site visits, and available studies:

- First, training and education efforts could increase awareness among government officials and the public about the impacts of climate change and available adaptation strategies. A variety of programs are trying to accomplish this goal, such as the Chesapeake Bay National Estuarine Research Reserve (partially funded by NOAA), which provides education and training on climate change to the public and local officials in Maryland.
- Second, actions to provide and interpret site-specific information could help officials understand the impacts of climate change at a scale that would enable them to respond. About 80 percent (147 of 183) of the respondents rated the "development of state and local climate change impact and vulnerability assessments" as very or extremely useful.
- Third, Congress and federal agencies could encourage adaptation by clarifying roles and responsibilities. About 71 percent (129 of 181) of the respondents rated the development of a national adaptation strategy as very or extremely useful. Furthermore, officials we spoke with at our site visits and officials who responded to our questionnaire said that a coordinated federal response would also demonstrate a federal commitment to adaptation.

Our past work on crosscutting issues suggests that governmentwide strategic planning can integrate activities that span a wide array of federal, state, and local entities.[6] As our report and others (such as the National Academy of Sciences and the Intergovernmental Panel on Climate Change) demonstrate, some communities and federal lands are already seeing the effects of climate change, and governments are beginning to respond. However, as our report also illustrates, the federal government's emerging adaptation activities are carried out in an ad hoc manner and are not well coordinated across federal agencies, let alone state and local governments. Multiple federal agencies, as well as state and local governments, will have to work together to address these challenges and implement new initiatives. Yet, our past work on collaboration among federal agencies suggests that they will face a range of barriers in doing so.[7] Top leadership involvement and clear lines of accountability are critical to overcoming natural resistance to change, marshalling needed resources, and building and maintaining the commitment to new ways of doing business. Given the complexity and potential magnitude of climate change and the lead time needed to adapt, preparing for these impacts now may reduce the need for far more costly steps in the decades to come.

Accordingly, our report released today recommends that the appropriate entities within the Executive Office of the President, such as CEQ and the Office of Science and Technology Policy, in consultation with relevant federal agencies, state and local governments, and key congressional committees of jurisdiction, develop a national strategic plan that will guide the nation's efforts to adapt to a changing climate. The plan should, among other things, (1) define federal priorities related to adaptation; (2) clarify roles, responsibilities, and working relationships among federal, state, and local governments; (3) identify mechanisms to increase the capacity of federal, state, and local agencies to incorporate information about current and potential climate change impacts into government decision making; (4) address how resources will be made available to implement the plan; and (5) build on and integrate ongoing federal planning efforts related to adaptation. CEQ generally agreed with the recommendation, noting that leadership and coordination is necessary within the federal government to ensure an effective and appropriate adaptation response and that such coordination would help to catalyze regional, state, and local activities.

Mr. Chairman, this concludes my statement. I would be pleased to respond to any questions you or other Members of the Committee may have.

End Notes

[1] Major greenhouse gases include carbon dioxide (CO_2); methane (CH_4); nitrous oxide (N_2O); and such synthetic gases as hydrofluorocarbons (HFC), perfluorocarbons (PFC), and sulfur hexafluoride (SF_6).

[2] GAO, *Climate Change Adaptation: Strategic Federal Planning Could Help Government Officials Make More Informed Decisions*, GAO-10-113 (Washington, D.C.: Oct. 7, 2009).

[3] For our questionnaire, 187 of 274 officials responded for a response rate of approximately 68 percent. Not all officials responded to every question.

[4] Secretarial Order No. 3289 (Sept. 14, 2009).

[5] Differences by level of government (federal, state, and local) that are reported are for illustrative purposes and may not be statistically different. We present selected examples where the difference between federal, state, or local responses is greater than 15 percent and the difference presents useful context for the overall results. There were other differences by level of government that are not presented in our report.

[6] GAO, *A Call For Stewardship: Enhancing the Federal Government's Ability to Address Key Fiscal and Other 21st Century Challenges*, GAO-08-93SP (Washington, D.C.: Dec. 17, 2007).

[7] GAO, *Results-Oriented Government: Practices That Can Help Enhance and Sustain Collaboration among Federal Agencies*, GAO-06-15 (Washington, D.C.: Oct. 21, 2005), and *Managing for Results: Barriers to Interagency Coordination*, GAO/GGD-00-106 (Washington, D.C.: Mar. 29, 2000).

In: Climate Change Adaptation
Editor: Elizabeth N. Brewster

ISBN: 978-1-61728-889-0
© 2010 Nova Science Publishers, Inc.

Chapter 4

TESTIMONY OF ERIC SCHWAAB, DEPUTY SECRETARY MARYLAND DEPARTMENT OF NATURAL RESOURCES BEFORE THE U.S. HOUSE OF REPRESENTATIVES SELECT COMMITTEE ON ENERGY INDEPENDENCE AND GLOBAL WARMING

October 22, 2009

Chairman Markey and distinguished members of the Select Committee, it is my pleasure to be here today to outline some of Maryland's successes in planning for climate change and to discuss with you the importance of developing a strategic national approach to adaptation.

Given our more than 4,000 miles of coastline and documented rate of sea level rise nearly twice that of the global average, Maryland has already begun to strategically plan for the impacts of climate change. In April 2007, Governor Martin O'Malley signed an Executive Order establishing the Maryland Climate Change Commission. Approximately a year after its formation, the Commission released Maryland's Climate Action Plan', setting forth a course of action to stem not only the drivers of climate change but also for how to adapt and respond to the inevitable consequences.

Historic tide-gauge records show that sea levels are rising along Maryland's coast and have increased one-foot within state waters over the last 100 years. We are currently expecting that sea level may rise at least twice as fast as it did during the 20th century, resulting in potentially 2.7 to 3.4 feet of rise by the year 2100. Such a rise will likely cause increased vulnerability to storm events, more frequent and severe coastal flooding, inundation of low-lying lands, submergence of tidal marshes, more shore erosion, salt-water intrusion, and higher water tables. While Maryland's entire coast will be impacted over the course of time, our state's low- lying coastal areas, to as well as those with large amounts of exposed shoreline are most at risk. The Chesapeake Bay is ranked the third most vulnerable region in the nation to the impact of sea level rise.

Confirming this fact is that the impact of sea level rise is already apparent - Maryland is currently losing approximately 580 acres per year to shore erosion; and alarmingly, thirteen Chesapeake Bay islands once mapped on nautical charts have already disappeared beneath the water's surface. In a 2008 report, the National Wildlife Federation estimated that approximately 400,000 acres of land on the Chesapeake's Eastern Shore could gradually be submerged." Maryland has thousands of miles of developed waterfront property along its coast, including many historic human settlements such as Smith Island. These coastal areas contain billions of dollars worth of public and private investments that will be adversely impacted by sea level rise and the intensification of coastal storm events.

A key component of Maryland's Climate Action Plan is the Comprehensive Strategy for Reducing Maryland's Vulnerability to Climate Change. Phase I of this Strategy sets forth the state's vision for protecting Maryland's future economic well-being, environmental heritage and public safety from the already inevitable impacts of climate change-induced sea level rise and coastal storms. The Strategy recommends a suite of 18 specific legislative, policy, and planning actions aimed at the reduction of impact to existing built environments & future growth and development; financial and economic impact avoidance; the protection of human health, safety and welfare; and the protection and restoration of the State's forests, wetlands and beaches as they inherently protect us from the impacts of climate change.

Implementation of the Adaptation Strategy is well underway. In 2008 Maryland passed two pieces of key legislation called for in the Strategy: The Living Shoreline Protection Act and amendments to the Chesapeake and Coastal Bays Critical Area Act. Both will reduce Maryland's vulnerability over time and protect natural resources from the impacts of sea level rise by restoring natural shoreline buffers such as grasses and wetlands and limiting new growth in vulnerable areas.

The work of the Maryland Commission on Climate Change drew national accolades for its focus on adaptation. The U.S. Climate Change Science

Program, Synthesis and Assessment Product 4.1, *Coastal Sensitivity to Sea Level Rise: A Focus on the Mid-Atlantic Region'* states that "Maryland has taken a proactive step towards addressing a growing a problem by committing to implementation of its sea level rise response strategy and increasing awareness and consideration of sea level rise issues in both public and governmental arenas."

Aside from sea level rise and coastal storms, Maryland is equally concerned about the likely consequences of a changing climate to the state's agriculture industry, forestry resources, fisheries resources, freshwater supply, aquatic and terrestrial ecosystems, and human health. For example, many marine living resources will likely experience changes in species composition and abundance with warming Fisheries managers will need to adapt management to account for shifts in productivity, variability and predictability of fish populations due to climate change.

In terms of water quality, a changing climate will have multiple and complex effects on the Chesapeake Bay as well as on Maryland's coastal bays and the nearshore ocean environment. Maryland and the other Bay states are taking aggressive action to accelerate Bay restoration efforts yet are concerned that rising sea levels and changes in precipitation patterns may make restoration more difficult to achieve. The Chesapeake Bay Program's Scientific and Technical Advisory Committee report, *Climate Change and the Chesapeake Bay*[6] and the Executive Order 13508, Draft Section 202(d) Report, *Chesapeake Bay Watershed Climate Change Impacts'*, both address this concern and collectively recommend the need for action on planning for adaptation at the regional, state and national level.

Recognizing the critical need to plan for such impacts as just described, the Maryland Commission on Climate Change has initiated development of Phase II of its Adaptation Strategy. This second phase is focused on addressing the impacts of increasing temperature, changes in precipitation patterns and increased storminess to six issue-based sectors: water resources, agriculture, aquatic and terrestrial ecosystems, forestry, agriculture, human health and transportation and land-use. Adaptation strategies for each sector are to be produced by June 2010.

Over the years, Maryland's coastal adaptation efforts have benefited from a variety of federal funding sources, most notably through the Coastal Zone Management Act (CZMA) as administered by the National Oceanic and Atmospheric Administration (NOAA). Maryland receives approximately $3 million annually from NOAA to implement these programs and has used approximately $250,000 of these yearly funds since the year 2000 to fund its sea level rise, coastal hazards and coastal climate change adaptation planning efforts. Thanks to this ongoing federal support, Maryland has a strong Coastal Zone Management and National Estuarine Research Reserve Program, both authorized under the CZMA. CZMA funds have supported climate change-related activities for research and data acquisition, as well as to expand technical, planning, and education activities needed to address key coastal climate change adaptation issues.

Our vulnerability to climate change will ultimately depend upon the magnitude of future impact, as well as how we as a society are able to cope and respond. In Maryland, however, we are continuing to invest, live, and actively manage lands and resources that we know with near certainty will be impacted by climate change. As a result, more and more of Maryland's people, property, public investments and natural resources, including vital fish and wildlife habitat, will soon be at risk.

If states and local governments do not adequately prepare for climate change, we may jeopardize at least a century of land and water conservation success across the country. The billions of dollars of investment in public lands and fish and wildlife habitat by federal and state agencies over the last 100 years is threatened by the anticipated pervasive impacts of climate change. We must protect the integrity of our investment in our nation's natural infrastructure to ensure our security. Federal, state and local governments must move beyond traditional planning and resource management practices and set a course for planning in anticipation of future change.

States are at the front lines of planning for climate change. Despite the absence of a national climate change adaptation program, States like Maryland are already undertaking significant strategic planning efforts. However, the efforts of states across the nation could be improved upon and assisted, by several key actions at the national level.

First, the key role of states, including the valuable research contribution of state academic institutions, in climate change adaptation planning must be clearly established and supported by federal programs

Next, to facilitate effective coastal adaptation the nation needs a clear national strategy for intergovernmental coordination on adaptation. This strategy should advocate an integrated national approach to natural resource adaptation that reflects meaningful coordination among the state and federal agencies. In the Chesapeake Bay region alone, at least three separate climate change adaptation strategies have been produced in the last year and half — all by different governmental organizations and all calling for enhanced intergovernmental coordination.

And finally, action at the federal level must provide dedicated funding for adaptation. Federal financial support is imperative to protect coastal communities, natural resources and the national interest from the impacts of climate change.

Along these same lines, some of the barriers to strategic coordination of adaption efforts across federal, state and local governments can be addressed by the following efforts:

- Reauthorize and strengthen the Coastal Zone Management Act. Strengthen the CZMA with authorization for climate change related activities; including funding to develop and implement a coastal adaptation plan that recognizes each state's individual needs while building into a proactive national strategy.
- Support creation of a permanent ocean trust fund. Revenues from this trust fund could be used by Maryland to address the impacts of climate change, including maintaining healthy, resilient coastal communities and economies; and protecting and restoring coastal ecosystems, habitats, waters, and unique resources. A potential source of revenue could be funds generated by greenhouse gas cap and trade programs.
- Improve awareness and understanding of the resources available to states and local governments. A key component of a national climate change adaptation program should be a new and stronger focus on intergovernmental coordination between federal, state and local agencies.
- Create a better system of observations at the national level - one that is reliably continuous, strategically targeted, and thoroughly integrated. There is generally insufficient monitoring of Maryland's climate, environmental conditions and resources to characterize their present state and variability. Reliable observations, interpreted with scientific understanding and innovative models, can dramatically reduce uncertainty about the path of climate change in Maryland and its consequences.
- Help states effectively respond to changes to aquatic and terrestrial ecosystems. Federal agencies should work closely with coastal states to assess impacts to coastal, marine and migratory fishery habitats and strategically target funds toward projects which will further adaptation. Coastal wetlands and bay islands are vital natural systems

in terms of the ecosystem services they provide in the form of clean water, clear air, storm buffers, and flood attenuation. Climate impacts such as drought, catastrophic fires and desiccation of wetlands will all result in releasing carbon currently sequestered in forests and wetlands. Functioning ecosystems sustain fish and wildlife and support associated fishing, hunting and wildlife- dependent recreation with an approximate national value of $76 billion per year — Maryland's portion of which is a tremendous asset to rural communities throughout the Chesapeake Bay region.

- Ensure federal adaptation funding for state and private forest lands. Climate change threatens the ability of the nation's forests - both public and private - to provide clean air and water, carbon sequestration, renewable energy and numerous other ecosystem services. Changes in precipitation, temperature, fire patterns, increased CO_2 concentrations, pest outbreaks and other climate change influences have the potential to transform forest ecosystems. Nearly two-thirds of the nation's forests are held in state and private ownership and will be essential in any wildlife and forest adaptation strategy. Funding adaptation activities on federal forests is essential, but only addresses the needs of a third of the nation's forests.

- Enhance smart growth programs and policies at the national level. Action is needed now to protect not only existing human settlements and infrastructure but also to ensure that we avoid future risk by restricting new growth and development in areas we already know are extremely vulnerable. Maryland is working to advocate smart growth practices as a means to accomplish this task and would advocate for the same level of effort at the national level. In the face of climate change, better land-use planning is imperative.

Preparing and planning for the consequences of climate change translates into more green jobs. Maryland's Governor, Martin O'Malley, established a goal of creating at least 100,000 green jobs by 2015. Adapting to climate change is one of the pillars of his Green Jobs Initiative. Numerous green jobs can be created through activities to support climate change adaptation, including marine contractors and landscape architects to design and install living shorelines; foresters to ensure sustainable forest management; biologists to address and remove invasive species, and habitat engineers to restore wetlands.

In conclusion, I want to highlight the need for government to lead by example. Federal, state and local government leadership is imperative if we are to combat and adapt to climate change. Maryland's state government is working to lead by example on the climate front by developing standards to guide the siting and design of state facilities and infrastructure in vulnerable coastal areas; working to reduce our footprint by sequestering carbon; and improving the efficiency of our vehicle fleet.

The issuance of the *Water Resource Policies and Authorities for Incorporating Sea-Level Change Considerations in Civil Works Programs'* by the U.S. Army Corps of Engineers in July 2009 represents a great instance of leading by example at the federal level. As does the Draft Strategic Plan of the U.S. Fish & Wildlife Service, *Rising to the Challenge, Responding to Accelerating Climate Change'*. I commend the federal government for such efforts to lead by example which set the stage for states to follow.

Thank you very much for your time in considering my testimony today.

REFERENCES

[1] Maryland Commission on Climate Change. 2008. *Maryland Climate Action Plan*. Maryland Department of Environment. Baltimore, Maryland.

[2] Glick, Patty, et.al.. 2008. *Sea-Level Rise and Coastal Habitats in the Chesapeake Bay Region*. National Wildlife Federation. Reston, Virginia.

[3] U.S. Climate Change Science Program and the Subcommittee on Global Change Research. 2009. *Coastal Sensitivity to Sea Level Rise: A Focus on the Mid-Atlantic Region*. Synthesis and Assessment Product 4.1.

[4] Pyke, C. & Najjar, R. et al. (2008). *Climate Change and the Chesapeake Bay*. Chesapeake Bay Program Scientific and Technical Advisory Committee. Annapolis, MD.

[5] Dept. of Commerce and Dept. of Interior. 2009. September 9, 2009 Draft Report on Chesapeake Bay Watershed Climate Change Impacts: A Draft Report fulfilling Section 202(d) of Executive Order 13508. Washington, D.C.

[6] I Department of the Army, U.S. Army Corps of Engineers. 2009. Water Resource Policies and Authorities for Incorporating Sea-Level Change Considerations in Civil Works Programs. Circular No. 1165-2-211.

[7] U.S. Fish & Wildlife Service. 2009. September 21, 2009 Draft Rising to the Challenge: Strategic Plan for Responding to Accelerating Climate Change. Reston, VA.

In: Climate Change Adaptation
Editor: Elizabeth N. Brewster

ISBN: 978-1-61728-889-0
© 2010 Nova Science Publishers, Inc.

Chapter 5

Testimony of Stephen Seidel, Vice President for Policy Analysis, Pew Center on Global Climate Change, Before the Select Committee on Energy Independence and Global Warming, Hearing on "The Federal Government's Role in Building Resilience to Climate Change"

Mr. Chairman, Mr. Sensenbrenner, members of the Select Committee, thank you for the opportunity to testify on the topic of what the federal government should be doing to adapt to climate change. My name is Stephen Seidel and I am Vice-President for Policy Analysis at the Pew Center on Global Climate Change.

Our Changing Climate

Responding to the risks of climate change represents one of the major challenges facing our nation and the global community. Most of the attention to date has appropriately been placed on actions to reduce emissions of greenhouse gases. This is obviously the first and best line of defense against the risks associated with global warming. But as our scientific understanding of climate change has improved, we also have come to realize that our past emissions have already begun to affect our current climate. Climate change isn't some distant concern that will impact our children or grandchildren. There is clear and convincing evidence that we have already experienced the following changes:

- U.S. temperatures have increased by more than 2 degrees F. over the past 50 years.
- Average global sea level has risen by 8 inches over the last century.
- The amount of rain falling in the heaviest downpours (the heaviest 1%) has increased by 20 percent over the last century.
- Arctic sea ice is declining dramatically - end of summer ice losses have averaged 11% per decade over the past three decades.[1]

The changes we've experienced to date are likely to increase dramatically over time. In fact, one of the unfortunate aspects of our climate system (due to built-in lags such as absorption of heat by the oceans) is that even if we could wave a magic wand and totally stop emissions of greenhouse gases immediately, global average temperatures would increase by another 1 degree F. If we continue on our current path and global greenhouse gas emissions continue to increase, temperatures would further rise, for a total increase on the order of 7-11 degrees F. by 2100

To reduce the damages associated with changes of this magnitude, two imperatives must be addressed:

1). We must take action to reduce greenhouse gas emissions to limit both the rate of climate change and the ultimate magnitude of that change.
2). We must take actions to minimize the costs associated with the unavoidable climate change that is already underway and will continue for many decades.

The second point is the focus of this hearing and a study that the Pew Center[2] of undertook to explore what the federal government should do to provide leadership to our nation in its effort to more effectively adapt to climate change.

OUR VULNERABILITY TO CLIMATE CHANGE

Climate is something we generally take for granted until it does something unexpected. Many key aspects of our economy are based on the critical assumption that our future climate will be similar to what we have experienced in the past. For example,

- Agriculture – what, where, and when we plant depends on temperature, length of growing season and water availability.
- Community development – what we build and where we locate structures and development depend on such factors as the availability of water, temperatures, risks of wildfires and coastal impacts.
- Energy development – many sources of electricity require large amounts of water for cooling, and different types of renewable energy depend critically on the availability of stream flows, sunlight or wind.
- Public health systems – are designed to anticipate and treat different types of diseases whose geographic ranges and seasonal occurrence may be influenced by climatic conditions.
- Emergency response systems – are designed around the likelihood and magnitude of extreme weather events (e.g., storms, floods, drought, and heat waves).
- National security – growing recognition among security experts that climate change, such as extreme weather events, scarcity of food, coastal flooding, etc. can contribute to increased tensions.
- Natural resources - the habitat for plants and animals, the viability of forests and the health of wetlands are all affected by temperature and the availability of water.

It should be clear that the impacts of changing our climate cut across a broad swath of our economy from food to energy production, to where and in what we live and how we travel, to the wellbeing of our natural resources and even to our national security. And that critical assumption - that future climatic conditions will be similar to the past - moves further and further away from reality with each ton of carbon dioxide we add to the atmosphere.

Damages from climate change are often discussed in terms of the impact that an average change in temperature or precipitation could cause. Yet we know from real life experience that the occurrence of extreme events (such as heat waves, floods, and intense storms) is what drives economic losses. We also know that one of the insidious aspects of climate change is that the number of extreme events is expected to increase dramatically. For example, under a scenario where emissions continue to grow uncontrolled, the number of days over 90 degrees in the Southern United States would increase from 60 per year to 150 per year by the end of the century.[3] With a one-half meter rise in sea level, the maximum level of flooding that New York City used to experience once every one hundred years would occur once every 25 years.

Role of the Federal Government

It is sometimes said that "all adaptation is local." This expression makes good sense in that climate impacts occur at a particular time and place and therefore are indeed local. Nonetheless, we believe that for our nation to build an economy more resilient to climate change, the federal government's role is critical for the following reasons:

1. Federally owned assets are at risk

The federal government owns 29% of all lands in the country. It owns 476,000 public structures including bridges, tunnels, and flood control structures that are valued at $723 billion. The Department of Defense alone has vast holdings many of which are in coastal areas. Naval bases are of course at sea level, but so are many Air Force bases and training bases such as Camps Pendleton in southern California and Leieune in coastal North Carolina. Many of our prized national parks are also vulnerable to the impacts of climate change and key features of some such as the Everglades and Glacier National Park could mostly disappear or be substantially changed. To properly manage these assets it is critical that the federal government understand the risks posed by climate change and the opportunities to adapt in a timely and cost-effective manner.

2. Federal guidelines, standards, and regulations are used across the economy

The federal government influences many decisions made by state and local governments and the private sector. The federal government is involved, directly or indirectly, in setting air and water pollution control regulations, in transportation and water infrastructure planning and design, and in design and siting of hydroelectric and other energy facilities. Other federal programs, such as the national flood and crop insurance programs, also play a major role in decisions that are affected by climate change.

3. Federal technical support is critical

The federal government provides critical information and technical support in areas related to climate and its impact. Weather information and hazardous conditions advisories are part of the daily fabric of our lives. The federal government's technical expertise is also made widely available through such mechanisms as the National Climatic Data Center, the Agricultural Cooperative Extension Service, the National Institute of Standards and Technology, federal land and forestry managers, and the Public Health Service.

Federal leadership in each of these areas is critical. If properly directed these and similar resources across the federal government can become important cogs in a national effort to adapt to climate change. Moreover, these same resources could play a significant role in both assisting state and local governments and the private sector in their adaptation activities.

We recommend a comprehensive review of federal activities aimed first at identifying assets, programs and activities most at risk from climate change and then making the necessary changes to enhance resiliency. Above all, we recommend that recognition of our changing climate be "mainstreamed" across all relevant federal programs. Nowhere should the federal government continue to assume that our future climate will be the same as the past.

Mainstreaming Adaptation

In analyzing how to structure a federal adaptation program it quickly became clear that one frequently used approach would not work. Adaptation is not the type of new issue where it would make sense to set up a new office or department and charge it with tackling the problem. It must be integrated into the everyday decisions of program managers across a wide spectrum of climate-related activities. Coastal zone managers must begin taking sea level rise into account when planning new development or shoreline protection. Agricultural agents must begin thinking about changes in growing seasons, temperatures, and water availability when deciding on seed selection or crop rotations. Land managers must consider fire risk changes resulting from shifts in precipitation, or damage to forests due to pest infestations (such as bark beetle infestations). Transportation planners must consider flood hazards when designing and locating new roads or bridges. None of these are new decisions, but each must be viewed with a new perspective - that future climate will be altered. Only by "mainstreaming" adaptation considerations across all relevant programs will our nation be in a position to meet the challenges of unavoidable climate change.

Based on a review of adaptation programs initiated by other countries and by state and local governments in the United States, we have developed the following recommendations.

1. *Federal Agency Strategic Plans*

We believe that a critical starting point is that each agency should develop its own strategic plan for what it needs to do to build greater resilience to climate change into its programs and mission. Agencies should begin by looking at their own programs because these will be the easiest to address and will also help them identify areas that need to be coordinated with other agencies or entities. Each agency's strategic plan should include the following:

- Identify and assess climate-sensitive assets, programs, policies, regulations and projects;
- Engage key stakeholders as part of the planning process;
- Identify barriers to incorporating climate change into agency decision-making and resource needs for implementation;
- Identify and develop priorities among the most vulnerable areas and response actions;
- Establish plans to monitor and evaluate implementation;
- Define areas requiring coordination with other agencies and partners; and
- Identify future research needs.

The good news is that several agencies have taken the first step down this path. In January 2009, the Secretary of Interior issued an order requiring bureaus and offices to "consider and analyze potential climate change impacts when undertaking long-range planning exercises, setting priorities for scientific research and investigations, developing multiyear management plans, and making major decisions regarding potential use of resources under the Department's purview." The order goes on to require that offices identify legal barriers, resource needs and recommended actions to respond to potential climatic impacts. In September 2009, this order was supplemented by the creation of Climate Change Response Councils and regional response centers to facilitate information sharing and response strategies across the Department. Within the Department of Interior, the U.S. Fish and Wildlife Service released its climate change strategy and five-year action plan in September of 2009. EPA's Office of Water has also has issued a strategic plan to address the impacts of climate change on its programs.

The White House, working through its Council on Environmental Quality, could play an important role in advancing the development of an effective federal program for adaptation. It took an important step in that direction recently as part of its Executive Order on "Federal Leadership in Environmental, Energy and Economic Performance." (October 5, 2009) This order requires each agency to develop an Agency Strategic Sustainability Performance Plan. As part of that plan each agency is required to

> "(i) evaluate agency climate-change risks and vulnerabilities to manage the effects of climate change on the agency's operations and mission in both the short and long term;"

This requirement could serve as a lynchpin for initiating an effective strategic planning process within agencies. Even before the executive order was issued, CEQ had been working with an interagency group identifying actions that could be taken to begin developing both agency and sector-specific strategic plans. These are encouraging initial signs of executive office leadership, but follow-up will be critical to ensure that agencies are committed to pursuing the internal engagement required for an effective planning and implementation process.

We fully recognize that in developing their strategic plans, agencies are likely to identify a number of key areas where program responsibility is shared with other agencies and with state and local entities. We believe that an important next step in the planning process is to identify areas where sector-specific plans are required. Such areas as water resources, land management, human health, ecosystem protection, and coastal protection are examples where multiagency efforts with strong stakeholder participation will be required. Finally, we believe that over time it would be useful and possible to combine agency and sector plans as key building blocks in the development of a national strategic plan. The national plan can help provide strategic direction, set priorities, and identify key milestones. This can best be done based on the content of more detailed plans (bottom-up) rather than be developed first (top down).

2. Creating a national climate service

A key requirement for adapting to climate change is the availability of information detailing what those changes are likely to be. In addition, technical support in how to use such information in decision making on adaptation will be critical. A national climate service would be the entity responsible for developing and communicating credible and actionable climate scenarios and projections for use in adaptation planning purposes. In every case study we examined, one of the first questions asked was what temperature change, sea level increase, or change in precipitation should we assume as the basis for our adaptation planning. Given the local scale at which these questions are being asked and the uncertainty about important aspects of predicting future climate change, particularly at that scale, providing useful information is not a trivial matter. Many state and local entities have turned to nearby university experts as a source for climate scenarios. This has worked well in many cases, but the idea of making useful and consistent climate information widely available on a national basis has attracted attention for many years.

The leading proponent of a national climate service has been the National Oceanic and Atmospheric Administration (NOAA). As a lead federal agency in developing climate observing systems, data analysis and predictions, NOAA has the scientific foundation upon which a national climate services could be constructed. It has undertaken an extensive process examining different ways to structure such an entity and has begun moving forward in its development.

In our review of how a climate services might be structured we divided the function into two key parts: the development and provision of the climate data and the support required by the user community (such as coastal managers, water planners, agricultural agents, and transportation planners) to effectively identify their needs and use the information provided. We found that NOAA could most effectively lead the first element but not necessarily the outreach and user community engagement.

To effectively engage the critical and diverse user communities, we believe a national climate services should involve other key federal agencies as sector working group leads. Figure 1 shows the proposed organizational structure. For example, the Department of Agriculture could lead a sector working group for the farm community and the Department of Interior could lead a sector working group on natural resources. The sector working groups would be responsible for fully engaging state and local entities, the private sector and other stakeholders in identifying the needs for information and decision support tools specific to their sector, in setting priorities and in communicating information to the sector.

3. Structuring a Federal Adaptation Program

Our analysis focused on how to integrate an adaptation program into other climate related program activities across the federal government. We examined the two programs that currently exist - the Global Change Research Program (GCRP) and the Climate Change Technology Program (CCTP). Both are established by statute and we would recommend that a national adaptation program also be established by legislation. Because of the desirability of executive office leadership, we recommend that the national adaptation program be chaired or co-chaired by CEQ or the Office of Science and Technology Policy. The program should be managed by a coordinating committee that is made up of senior policy officials from each of the relevant agencies. We also recommend the creation of a small program office (along the lines of the office created under GCRP) to coordinate the agency and sector strategic planning activities.

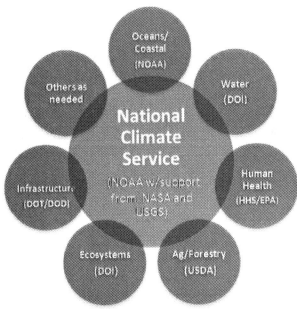

Figure 1. Proposed National Climate Service Sector Working Group Structure

4. Mandating Adaptation Considerations under the National Environment Policy Act (NEPA)

To ensure that adaptation is considered in all major federal actions, we recommend issuing clarifying regulations under N EPA. These regulations would make it clear that climate change needs to be considered in the planning stage of any major federal action. CEQ is responsible for NEPA's implementation, while EPA's Office of Federal Activities reviews environmental impact statements. We suggest establishing an interagency working group to prepare the proposed regulatory changes and to develop guidance for agencies in preparing EISs.

Adaptation Provisions in the American Clean Energy and Security Act (H.R. 2454)

The American Clean Energy and Security (ACES) Act passed by the House of Representatives in June of this year contains several provisions to address the issue of domestic climate change adaptation. While we we were pleased to see an adaptation section in the bill, we believe there are at least two important improvements that should be considered. First, the development of adaptation strategic plans for all relevant federal agencies is a key component of improving our nation's resiliency to climate change. As currently written, the ACES Act only contains provisions for natural resource agency adaptation plans and a public health strategic plan. Second, although the ACES Act does contain provisions establishing a national climate service within NOAA, we would recommend a structure similar to the one outlined above that both provides for a central role for NOAA, but also more effectively engages other key agencies as sector working group leads.

Conclusion

In conclusion, I would like to thank the Chairman, Mr. Sensenbrenner, and the members of the Select Committee for their time and attention to the important matter of furthering the U.S. government's efforts to address climate change adaptation.

End Notes

[1] Global Change Impacts in the United States, Thomas Karl, Jerry M. Melillo, and Thomas C. Pederson, (eds.) Cambridge University Press, 2009

[2] The Pew Center on Global Climate Change will be issuing a report before the end of the year detailing its analysis of the federal role in adaptation. Supporting the Pew Center in this research has been Stratus Consulting and Terri Cruce, an independent contractor.

[3] Karl, Melillo, and Pederson (eds.), Climate Change Impacts in the United States, pg. 34.

In: Climate Change Adaptation
Editor: Elizabeth N. Brewster

ISBN: 978-1-61728-889-0
© 2010 Nova Science Publishers, Inc.

Chapter 6

TESTIMONY OF DR. KENNETH P. GREEN BEFORE THE HOUSE SELECT COMMITTEE ON ENERGY INDEPENDENCE AND GLOBAL WARMING HEARING "BUILDING U.S. RESILIENCE TO GLOBAL WARMING IMPACTS"

Chairman Markey, Congressman Sensenbrenner, Members of the Committee:

Thank you for inviting me to testify today on this important topic. Along with these remarks, I have submitted, for the record, a policy study that I recently completed, entitled "Climate Change: The Resilience Option." My testimony here today represents my personal views, and should not be construed as the official position of the American Enterprise Institute, or any other persons or organizations.

Before I begin my remarks, I always like to list my three B's, my background, biases, and beliefs.

As to background, I am a biologist and environmental scientist by training, an economist by exposure, and a policy analyst by vocation: I've spent the last 15 years analyzing environmental policy in think tanks in the U.S. and Canada.

My biases are for solving environmental problems, whenever possible, with instruments that maximize freedom, opportunity, enterprise, and personal responsibility. Thus, I strongly favor true market-based remedies for environmental problems over command-and-control regulation. (I will observe here that cap-and-trade legislation is not truly market-based, as government sets a limit on emissions, rather than allowing a market to determine that level. Cap-and-trade is more akin to rationing than it is to markets).

Finally, my beliefs are based on reading the scientific literature as well as the IPCC climate science reports, and I believe that while greenhouse gases do retain heat in the atmosphere (making Earth habitable), the heat-retention capability of additional anthropogenic greenhouse gases is modest. I do not believe in predictive climate models, or most other forms of forecasting other than simple extrapolation for very modest periods of time.

That being said, I do believe that climate science has taught us something important that merits action. We have learned the Earth's climate is not the slow-moving system we thought it was. Rather, the climate is prone to sharp shifts into cooler and warmer conditions that can depart significantly from "average" temperatures for decades at a time. Acting to enhance climate resilience is an important task.

So, to the issue at hand: how can we best build U.S. resilience to global warming impacts?

First, I believe that we should shift our focus from mitigation of greenhouse gas emissions toward an adaptation agenda. We do not, at present, have the technologies needed to significantly curb greenhouse gas emissions without causing massive economic disruption, and without preventing the developing countries from developing, and lifting their billions of people out of squalor and poverty. The money and attention that we are spending on mitigation efforts is largely wasted – even if we shut the U.S. and the EU off completely, the trajectory of emissions from China and India will negate the environmental benefit of our self-sacrifice completely in only a few years. All that jacking up energy costs will do is deprive of us economic productivity which is the ultimate wellspring of our resilience and well-being.

Second, I believe that we should stop making things worse. That is, we should remove the misguided incentives that lead people to live in climatically fragile areas such as the water's edge, drought-prone locations, flood-prone locations, and so on.

At present, our federal and state governments exacerbate this risk-taking by acting as the insurer of last resort. When people who live at water's edge or in a flood plain are hit by storms or floods, governments intervene not only to rescue them and their property if possible, but then to provide rebuilding funds to let the people build right back where they are at risk. We are currently doing this in New Orleans, where people are re-building in an area that is still at risk from storm surges and levee failure. Undoubtedly, we'll do this in California, putting people right back into fire-prone areas they were burned out of last year.

As Charles Perrow observes in his book *Our Next Catastrophe*: "State-mandated pools have been established to serve as a market of last resort for those unable to get insurance, but the premiums are low and thus these have the perverse effect of subsidizing those who choose to live in risky areas and imposing excess costs on people living elsewhere. In addition, the private insurers are liable for the net losses of these pools, on a market-share basis. The more insurance they sell, the larger their liability for the uninsured. Naturally, they are inclined to stop writing policies where there may be catastrophic losses. The Florida and California coastlines are very desirable places to live and their populations have grown rapidly, but these handsome lifestyles are subsidized by residents living in the less desirable inland areas in the state, and, to some limited extent, by everyone in the nation."

Programs that subsidize climatic risk-taking should be phased out as quickly as possible, in favor of fully-priced insurance regimes. Rebuilding after disasters in climatically fragile areas should be discouraged. Eliminating risk subsidies would show people some of the true cost of living in climatically risky areas, and would, over time, lead them to move to climatically safer places where they can afford to insure their property and safety.

Third, we must look to our infrastructure. Another government action that leads pe ople to live in harm's way is the failure to build and price infrastructure so that it is both sustainable, and resilient to change. Governments build highways, but generally without a pricing mechanism. Thus, no revenue stream is created to allow, for example, for the highway to be elevated if local flooding becomes a problem. There is also no price signal relayed to the users of the highway that reflects the climatic risk that their transportation system faces. The same is true of freshwater infrastructure, wastewater infrastructure, electricity, and other infrastructure. Politicians love cutting ribbons on new "free" infrastructure. They're less prone toward having the cost of that infrastructure show up in terms of tolls or user fees.

Establishing market pricing of infrastructure would quickly steer people away from climatically fragile areas, dramatically reducing the costs of dealing with climate variability.

For example, let's consider our electricity supply. As long as governments distort the prices consumers pay for energy with subsidies, fuel mandates, renewable power mandates, and the like, electricity markets cannot effectively adapt to changing climatic conditions. If electricity markets were fully deregulated, and if full costs were passed onto consumers, price signals would be created for the electricity provider in terms of expanding or decreasing capacity and for the consumer in terms of the real cost of living in an environment subject to energy-consuming heat waves (or cold snaps). Privatization would create incentives for electricity conservation and for the acquisition of energy-efficient appliances and devices without any need for specific governmental efficiency standards. Further, electric companies would be driven to connect with one another to ensure reliability to their customers rather than doing the minimum possible to satisfy regulators.

And consider our water supply. Full pricing of water and full privatization of the water supply, drinking water plants, and wastewater treatment plants would ameliorate many climatic risks incrementally over time, including flooding, seawater intrusion, and coastal and river pollution from storm runoff. Charging the full price for water, from supply to disposal, would create a price signal for consumers regarding the real risks they face living in hydrologically sensitive areas and create incentives for conservation while producing a revenue stream to allow for expanded capability or the securing of alternative supplies. At some point, again, high prices could simply lead people to move away from areas that are hydrologically costly, such as cities dependent on a single winter snow pack that shrinks or a single major river that suffers reduced flow.

Finally, I would suggest that we trust in resilience, but tie up our camel. In the event that climate change does tend toward higher estimates put forward by the United Nations and other groups, it is reasonable to consider insurance options that might help deal with such climate changes. Such options might include government investment in geoengineering research, investment in research and development to advance technologies allowing the removal of greenhouse gases from the atmosphere

Climate variability, whether natural or man-made, does pose significant challenges to the health of our population, the maintenance of our infrastructure, and to our economic growth. Taking steps to make our society resilient in the face of climate variability is an important endeavor, and I applaud your hearing on the matter today.

Thank you for allowing me to speak to you today on this timely and important issue.

In: Climate Change Adaptation
Editor: Elizabeth N. Brewster

ISBN: 978-1-61728-889-0
© 2010 Nova Science Publishers, Inc.

Chapter 7

COMPARISON OF CLIMATE CHANGE ADAPTATION PROVISIONS IN S. 1733 AND H.R. 2454

Melissa D. Ho

SUMMARY

This report summarizes and compares climate change adaptation-related provisions in the American Clean Energy and Security Act of 2009 (H.R. 2454) and the Clean Energy, Jobs, and Power Act (S. 1733). H.R. 2454 was introduced by Representatives Waxman and Markey and passed the House on June 26, 2009. S. 1733 was introduced to the Senate by Senators Boxer and Kerry and, after subsequent revisions made in the form of a manager's substitution amendment, was reported out of the Senate Environment and Public Works Committee on November 5, 2009.

Adaptation measures aim to improve an individual's or institution's ability to cope with or avoid harmful impacts of climate change, and to take advantage of potential beneficial ones. Both H.R. 2454 and S. 1733 include adaptation provisions that (1) seek to better assess the impacts of climate change and variability that are occurring now and in the future; and (2) support adaptation activities related to climate change, both domestically and internationally.

Overall, while the two bills would authorize similar adaptation programs, they differ somewhat in scope and emphasis, and they also differ in the distribution of emission allowance allocations over time. Both bills contain provisions that address international climate change adaptation; domestic climate change adaptation programs, including the U.S. Global Change Research Program (USGCRP), the National Climate Service, and state and tribal programs; public health; and natural resources adaptation. S. 1733 includes five additional provisions not provided for in the House bill that deal with drinking water utilities; water system mitigation and adaptation partnerships; flood control, protection, prevention, and response; wildfire; and coastal Great Lakes states' adaptation.

Neither the Senate-reported bill (S. 1733) nor the House-passed bill (H.R. 2454) contains a process at the federal level for developing and implementing a national strategic plan to address the full range of sectors expected to be affected by climate change. Neither bill includes provisions that explicitly address adaptation in major sectors such as transportation and energy infrastructure, or agriculture.

Another difference between S. 1733 and H.R. 2454 is the distribution of allowance allocations over time, and the subsequent availability of the amounts credited to certain funds. The relative distribution of allowances to adaptation-related activities is slightly higher in the House bill than in the Senate bill, and the difference increases over time, but the actual amounts of revenue generated would be contingent on the number and price of emission allowances. The Senate bill provides that funds for many adaptation-related provisions, such as for natural resources and public health, are made available "without further appropriations." In contrast, the analogous provisions in the House bill provide that the funds would become available only by subsequent appropriations.

A side-by-side table is included in an appendix to the report that compares adaptation-related provisions in H.R. 2454 and S. 1733.

INTRODUCTION

Congress is currently considering major legislation related to climate change. Climate change responses have typically been categorized into two broad types: mitigation and adaptation. Mitigation measures attempt to slow down the occurrence of climate change by, for example, reducing greenhouse gas emissions. Adaptation measures, on the other hand, aim to improve an individual or institution's ability to cope with or avoid harmful impacts of climate change, and to take advantage of potential beneficial ones. While much attention has been paid to mitigation efforts, a growing focus on current impacts of climate change has led to specific provisions in several bills that would increase research on and programmatic attention to possible options for adaptation.[1] Climate change mitigation and adaptation activities are not mutually exclusive, and in most cases can actually be complementary. Because the extent of climate change impacts upon different ecosystems, regions, and sectors of the economy will depend not only on the sensitivity of those systems to climate change, but also on the systems' ability to adapt to climate change, both types of activities are considered by many to be an essential part of a comprehensive approach to dealing with the impacts of climate change.

The American Clean Energy and Security Act of 2009 (H.R. 2454) passed the House on June 26, 2009. The Senate Environment and Public Works (EPW) Committee approved the Clean Energy, Jobs, and Power Act (S. 1733) on November 5, 2009.[2] Both H.R. 2454 and S. 1733 would establish a cap-and-trade system to regulate greenhouse gas emissions, and address energy topics including energy efficiency and renewable energy. Both bills also include adaptation provisions that (1) seek to better assess the impacts of climate change and variability that are occurring now and in the future; and (2) support adaptation activities related to climate change, both domestically and internationally.

This report summarizes and compares the adaptation-related provisions in H.R. 2454 and S. 1733. A side-by-side table in an Appendix to the report compares relevant provisions related to climate change adaptation in both bills. The provisions are grouped into the following headings:

- International Climate Change Adaptation
- Domestic Climate Change Adaptation (including the National Climate Change Adaptation Program and the National Climate Services Program)
- State and Tribal Programs
- Public Health
- Natural Resources Adaptation
- Other Climate Change Adaptation Programs, including Water Resources (in S. 1733 only)

CLIMATE CHANGE AND ADAPTATION

Importance of Adaptation

Climate-related changes have been observed in the United States and globally. A recent report by the U.S. Global Change Research Program (USGCRP) provided scientific documentation of the impacts of climate change already occurring in the United States.[3] The report analyzed different sectors and regions of the United States and concluded that climate disruption causes a wide range of damaging impacts in the United States currently, and that these impacts will continue to intensify, depending on the region. The report also found that population growth and increased use of resources will limit the ability of society and natural systems to adapt successfully. Specific findings include increased:

- stress on water resources, which will amplify regional droughts and reduce water supply, especially in regions dependent on western mountain snowpack;
- risk for coastal settlements, infrastructure, and ecosystems from sea-level rise and more intense hurricanes and storm surges;
- numbers of wildfires and areas of forest adversely affected or destroyed by pest outbreaks linked to warming;
- threats to human health related to heat waves, poor air quality, and insect-borne diseases;
- challenges to crop and livestock production due to increasing stress on water resources, increasing temperatures, increasing outbreaks of pests and diseases, and the need for new management practices;
- stress of population growth and overuse of resources, which will limit the ability of society and natural systems to adapt successfully.

The Intergovernmental Panel on Climate Change (IPCC) stated in its Fourth Assessment Report that "adaptation will be necessary to address impacts resulting from the warming which is already unavoidable due to past emissions."[4] The panel concluded that many industrial sectors and the natural environment, including agriculture, forestry, water resources, human health, coastal settlements, and natural ecosystems, will need to adapt to a changing climate or possibly face diminished productivity, functioning, and health.

Adaptation can include a wide range of activities. For agriculture, examples of adaptation can include farmers changing management practices—for example, altering their planting dates and irrigation scheduling—or farmers switching to different crop varieties altogether, in response to changing temperature and rainfall regimes. For coastal regions, strategies to prevent damage from climate change and rising sea levels can include improving shoreline protection measures—for example, installing dikes, levies, other structures, and beach vegetation—or can result in companies relocating key business centers away from coastal areas vulnerable to inundation and hurricanes. The costs of implementing adaptation measures are generally considered in relation to the value of the assets protected, in order to assess the net benefit of adaptation investments.

Current Status of Public Action

While adaptation as an approach for dealing with the impacts of climate change is gaining increasing attention, few examples exist of concrete actions or strategies dealing with adaptation across different levels of government. According to a recent report by the National Research Council (NRC), individuals and institutions are unprepared both conceptually and practically for meeting the challenges and opportunities that climate change presents.[5] Similarly, at a recent hearing of the House Select Committee on Energy Independence and Global Warming, experts testified that current U.S. adaptation efforts are largely ad hoc, uncoordinated, underfunded, and lacking the information needed to make critical decisions.[6] Specifically, testimony from the Government Accountability Office (GAO), based on its recently released report on nationwide climate change adaptation efforts, concluded that adaptation efforts are often constrained by a lack of site-specific data, such as local projections of expected changes, and by a lack of clear roles and responsibilities among federal, state, and local agencies.[7] The NRC report included recommendations to bolster the capacity of federal programs in the area of climate science and information; to strengthen research on adaptation, mitigation, and vulnerability; to initiate a periodic national assessment of climate impacts and responses; and to routinely provide policymakers and the public with the relevant scientific information, tools, and forecasts to make better-informed decisions.

Adaptation initiatives are starting to gain traction at the federal and state levels.[8] For instance, the Department of Interior (DOI) recently launched an internal agency initiative to develop a coordinated strategy to address current and future impacts of climate change.[9] The DOI initiative, which was established through secretarial order,[10] establishes a framework through which Interior bureaus will coordinate climate change science and resource management strategies. Also, the National Oceanic and Atmospheric Administration's (NOAA's) Regional Integrated Sciences and Assessments has a program that supports research to meet the adaptation-related information needs of local decision-makers. While federal agencies are beginning to recognize the need to adapt to climate change, there is still a general lack of strategic coordination across agencies, and most efforts to adapt to potential climate change impacts are preliminary.

Some states have begun to make progress on adaptation independently and through partnerships with other entities, such as academic institutions. The state of California recently developed and released a draft California Climate Adaptation Strategy.[11] This is the first example of a strategic, operational plan for collaborative action by state agencies to adapt to impacts of global climate disruption and sea-level rise. Maryland has also begun a strategic planning process to better understand the impacts of climate change on the state's economy and natural resources, especially the Chesapeake Bay region, and to coordinate state efforts.

In devising these strategic plans for adaptation, states are often calling for more resources, leadership, and coordination from the federal government. Specifically, state agencies such as those in California and Maryland are advocating for:

- more federal support for state research programs that generate locally relevant data and information related to potential impacts of climate change;
- an integrated national intergovernmental strategy on adaptation that is coordinated among relevant state and federal agencies, and is multidisciplinary and inclusive of other sectors such as transportation, energy, agriculture, forestry, water resources, and utilities;
- more dedicated federal funding for adaptation, to carry out programs to protect coastal communities, natural resources, and the national interest from the impacts of climate change.

Some are skeptical of implementing wide-scale adaptation measures and argue that adaptation activities should not be comprehensively pursued because attention and resources will detract from mitigation efforts. Others think that adaptation activities are just another way for various interest groups and sectors to seek government subsidies for activities they would already otherwise be doing.

OVERVIEW OF ADAPTATION PROVISIONS IN S. 1733 (AS REPORTED BY THE SENATE EPW COMMITTEE) VS. H.R. 2454 (AS PASSED BY THE HOUSE)

This report summarizes and compares the adaptation provisions in S. 1733, as reported by the Senate Environment and Public Works (EPW) Committee on November 5, 2009, and H.R. 2454, as passed by the House on June 26, 2009. Overall, while the two bills would authorize similar adaptation programs, they differ somewhat in scope and emphasis, and they also differ in the distribution of emission allowance allocations, which in effect provide monetary resources for specified programs and activities.[12] Both bills contain provisions that address:

- international climate change adaptation;
- domestic climate change adaptation programs, including the National Climate Change Program and the National Climate Service;
- state and tribal programs;
- public health; and
- natural resources adaptation.

S. 1733 contains five additional provisions (not contained in the House bill) that deal with:

- drinking water utilities;
- water system mitigation and adaptation partnerships;
- flood control, protection, prevention, and response;
- wildfire; and
- coastal Great Lakes state adaptation.

Neither the Senate-reported bill (S. 1733) nor the House-passed bill (H.R. 2454) contains a process at the federal level for developing and implementing a national strategic plan to address the full range of sectors expected to be affected by climate change. Neither bill includes explicit provisions that address adaptation in major sectors such as transportation and energy infrastructure, or agriculture, although these activities are allowable under state programs for climate adaptation that are provided for in both bills.

It should be noted that while forestry and agriculture are considered extensively in S. 1733 and H.R. 2454 with regard to supplemental emissions reductions, set-asides and allowances, and carbon offsets, these considerations are not specifically related to adaptation. Depending on the nature of the implementation, these programs could potentially assist in forest and agriculture adaptation to climate change. However, they have not been included in this report because emissions mitigation is their primary purpose (not adaptation), and adaptation is not necessarily a consideration in their implementation.

Allowance Allocations for Adaptation-Related Activities

Although there are significant differences in how the overall emission allowances are distributed, both bills would allocate allowances or auction revenues to fund various adaptation activities.[13] Table 1 provides an overview of the emission allowances allocated to adaptation-related activities for 2016 and 2030, given as a percentage of total allowances for both bills.

One significant difference between the two pieces of legislation is the distribution of allowances and proceeds from auction allocations and the subsequent availability of the amounts credited to certain funds. In the Senate bill, several of the adaptation-related provisions provide that the amounts in the funds are automatically available to be obligated (i.e., spent), "without further appropriation," for specified purposes, programs, and activities. In contrast, the analogous adaptation provisions in the House bill provide that the amounts in the funds would become available only by subsequent appropriations. That is, the amounts would not be available automatically, but instead would need to be provided in subsequent appropriations acts.

H.R. 2454 both allocates allowances directly and creates several funds for the allocation of proceeds from the sale of allowances. It generally allocates allowances to states and tribes, and proceeds from the auction of allowances to federal government agencies. Authorizations are subject to future appropriations. For the adaptation provisions, the House relies on hortatory language, such as that found in Section 480(b), to support full appropriations for certain natural resources programs: "... such sums as are deposited in the Natural Resources Climate Change Fund, and the amounts appropriated for subsection (c) shall be no less than the total estimated annual deposits in the Natural Resources Climate Change Adaptation Fund."

While the allocations of allowances and of the proceeds from auctions are distributed similarly by S. 1733, in the Senate bill, in all cases related to adaptation, the auction proceeds for programs or funds would be automatically available to be obligated (i.e., spent) "without further appropriation."[14] The provision of funding "without further appropriation" might be controversial. Comparable language in provisions of the House bill (none related to adaptation programs, however) has been scored by the Congressional Budget Office (CBO) as mandatory spending.[15] Even with mandatory funding, the ultimate funding of specific programs and activities will in many cases be determined by agency, state, and/or tribal decision-makers.

Table 1. Adaptation Allowances in S. 1733 vs. H.R. 2454.

	S. 1733 (EPW-reported bill) % of total allowances		H.R. 2454 (House-passed bill) % of total allowances	
	2016	2030	2016	2030
International Adaptation[a]	1.10	3.71	1.0	4.0
State and Tribal Adaptation[b]	0.43	1.62	0.90	3.9
Public Health	0.09	0.07	0.10	0.1
Natural Resource Adaptation[c]	0.87	2.97	1.01	4.0

Source: CRS analysis of S. 1733 (as reported by Senate EPW Committee) and H.R. 2454 (as passed by the House).

Notes: Percentages reflect the share of total allowances less those for the Strategic Reserve (H.R. 2454) or the Market Stability Reserve (S. 1733). The amounts of revenue would be contingent on the value of emissions allowances over time.

A. For S. 1733, this includes both off-the-top allowances and direct allowances for activities related to international adaptation.

B. Both bills include the establishment of a State Climate Change Response (SCCR) Fund in each state, which could be used to fund state and local government programs for greenhouse gas reduction and climate adaptation. Specifically concerning adaptation, funds are for state-administered grant programs related to transportation; water systems mitigation and adaptation partnerships; flood control and response; agriculture; and other activities. H.R. 2454 does not include several of the water resource provisions.

C. Natural resource adaptation includes both direct allowances and allowances obtained by auction.

International Adaptation

Developing countries, especially those that are least developed, and the poorest communities, are the most vulnerable to the impacts of climate change. In these vulnerable countries and communities, the impacts of climate change can pose a direct threat to people's very survival. Specific impacts highlighted by the Fourth Assessment Report of the Intergovernmental Panel on Climate Change (IPCC 2007)[16] include the following.

> By 2020, yields from the 93% of crop production in Africa that is rain-fed could be reduced by up to 50%.
>
> - Worldwide, approximately 20%-30% of plant and animal species are likely to be at increased risk of extinction if increases in global average temperature exceed 1.5°C -2.5°C.
> - Widespread melting of glaciers and snow cover will reduce melt water from major mountain ranges (e.g., Hindu Kush, Himalaya, Andes), where more than 1 billion people currently live.
> - Displacement of an estimated 200 million people due to sudden climate-related disasters is projected by 2050; it is estimated that in 2008 more than 20 million people were displaced by sudden climate-related disasters.
> - Increased adverse health impacts and mortality will result from higher frequency and intensity of climate-related diseases such as heat stroke, malaria, dengue, and diarrhea.

International assistance for adaptation, especially to help the most vulnerable developing countries, is one of the major commitments of industrialized countries under the United Nations Framework Convention on Climate Change (UNFCCC), to which the United States is a party. Adaptation assistance is also one of the major issues under negotiation in an effort to reach agreement in Copenhagen in December 2009 on international cooperation to address climate change beyond the year 2012.

Many have asserted that current overseas development aid (ODA) is insufficient to cover the adaptation needs of developing countries. A variety of international institutions and nongovernmental organizations have tried to estimate the costs of adaptation for developing countries and the associated needs for public funding. Figures range from $4 billion to several hundreds of billions of dollars annually by the year 2030, where definitions and scope of adaptation activities often account for many of the differences in funding requirements.[17] The World Bank, in an updated study from September 2009, estimates the average annual adaptation costs from 2010 to 2050 to be between $75 billion to $100 billion annually,[18] while EU leaders agreed in October 2009 that developing nations would need $150 billion annually by 2020 to tackle climate change and to deal with its consequences.[19] Estimates for climate change adaptation by sector made by the UNFCCC are given in Table 2.

Table 2. Climate Change Adaptation Cost Estimates by Sector Needed by 2030 (billion dollars per year normalized for 2009).

Sector	Global Cost	Developed Countries	Developing Countries
Agriculture	14	7	7
Coastal Zones	11	7	4
Human Health	5	not estimated	5
Infrastructure	8-130	6-88	2-41
Water	11	2	9
Total	49-171	22-105	27-66

Source: UNFCC (2007).

Much of the language in the House and Senate bills is identical, but there are several differences regarding programs to support international adaptation to climate change. Both bills establish an International Climate Change Adaptation Program, but S. 1733 would insert "and Global Security" into the title and makes clear that adaptation assistance should protect and promote U.S. interests. In H.R. 2454, Section 495 provides explicit authority for a variety of activities and aid eligible for support, including research, planning, investments, and capacity-building, among others. S. 1733 does not include a comparable specific list of eligible activities.

Both bills direct the Secretary of State or other designee of the President to distribute funding for international climate adaptation bilaterally or multilaterally. However, H.R. 2454 requires that 40% to 60% of funding go to multilateral funds or international institutions that meet given eligibility requirements. H.R. 2454 also specifies that no more than 10% of bilateral assistance may go to any one country; S. 1733 would not set such limits. The House bill also includes language directing that resources provided to this program must supplement, not supplant, other federal, state, or local resources that would similarly support international adaptation activities (i.e., requiring "additionality" of adaptation assistance). H.R. 2454 also does not explicitly provide for bilateral programs in other agencies that may have capacity-building, technological, financing, or other expertise that could be effective in assisting adapting to climate change.

While H.R. 2454 gives responsibilities for oversight of funding distributions to the Secretary of State (or other presidential designee) and the Administrator of the U.S. Agency for International Development (USAID), S. 1733 would give that authority to a Strategic Interagency Board on International Climate Investment.[20]

Domestic Adaptation

Improving adaptation in the United States to climate variability and change could include the following modified or new activities:

- climate observation and forecast services, such as season to interannual predictions of weather, or multi-decadal forecasts of temperature, precipitation, and other climate parameters;
- research and analysis on climate change impacts, including the identification of potential risks, benefits, and options for adaptation;
- development of vulnerability assessments and adaptation strategies within and across sectors, localities, states, agencies, and sectors;
- incorporation of climate variability and change into infrastructure planning and operating procedures;
- development of a comprehensive climate adaptation strategy that includes crosssectoral and interagency strategies and plans;
- testing and demonstration of adaptation measures; and
- evaluation, training, and information-sharing of successful programs and experiences related to climate adaptation.

In general, S. 1733, as reported by the Senate EPW Committee, and H.R. 2454, as passed by the House, reflect similar but not identical approaches and identified needs regarding climate change adaptation. Both bills would expand federal efforts to address adaptation to climate change, although the federal role is limited in different ways. The differences between the House and Senate bills reflect differences in priorities, and in determining which entities should be responsible for developing and implementing adaptation strategies. Neither of the bills is comprehensive in terms of authorizing an overarching strategy across sectors and levels of government.

Both bills would establish national "climate services" to develop observational data, climate modeling, and access to information for federal, state, local, and private decision-makers, to help them develop and execute adaptation strategies. Both bills apparently place primary responsibility in the states and Indian tribes for developing most strategies and plans for domestic adaptation, financially supported by sales of federal greenhouse gas emission allowances. S. 1733 also would require states to provide a part of their funding to localities for climate change adaptation. Both the House and Senate bills supplement the state responsibilities with requirements to establish sector-specific adaptation plans and activities at the federal level, emphasizing adaptation to protect public health and natural resources.

S. 1733 and H.R. 2454 provide authorities for domestic adaptation in three major categories:

- National Climate Change Adaptation Program;
- National Climate Service; and
- state and tribal adaptation programs.

National Climate Change Adaptation Program

At the national level, both H.R. 2454 and S. 1733 authorize expansion of the federal role in adaptation to climate change, although neither of the bills calls for comprehensive assessment and strategy to coordinate across levels of government. S. 1733, however, would establish a new national adaptation program. The Senate approach, as outlined in S. 1733, would establish a broad federal authority by directing the President to establish a National Climate Change Adaptation Program to increase "the overall effectiveness of Federal climate change adaptation efforts." The wording leaves wide discretion to the President, although it is seemingly limited to the interests and activities of federal agencies. For example, S. 1733 does not explicitly authorize that this program work with states, localities, and the private sector on cross-cutting strategies or to coordinate among different entities and stakeholders.

H.R. 2454 does not explicitly provide similar broad authority to the President to establish a national climate change adaptation program (although some might argue that the President already has such authority). While earlier versions of the House bill also would have established a comprehensive federal adaptation program and strategy, these provisions were not included in the version of H.R. 2454 passed by the House. H.R. 2454, as passed by the House, expands the focus of the U.S. Global Change Research Program[21] (USGCRP) to include climate change adaptation, vulnerability assessments, and policy analysis. While the coordinating committee for the USGCRP would expand beyond the current science agencies and research programs to include agencies representing sectors that have a stake in adapting to climate change, the implication of the H.R. 2454 language is that federal adaptation efforts remain primarily a research, and not a programmatic, effort. These and other USGCRP efforts would be led by the White House Office of Science and Technology Policy (OSTP).

Although S. 1733 would establish a National Climate Change Adaptation Program, it does not provide language authorizing funding or allocating emission allowances to the program. (Funding authorizations and allowance allocations are provided for other adaptation provisions in both bills.) H.R. 2454 increases authorization for interagency coordination of the USGCRP (not just for adaptation) to $10 million annually, approximately doubling recent expenditures.

National Climate Service

Many scientists and decision-makers agree that in order to adapt effectively to climate change, individuals and institutions need more accurate climate data and information that is specific to their locations and concerns. There is less agreement on the appropriate authorities, the scope of federal programs, and how federal programs should be structured and coordinated to implement plans and activities.

Both S. 1733 and H.R. 2454 would establish national programs to develop and provide access to information to assist decision-makers in plannning for adaptation to climate change. The National Climate Service in H.R. 2454 would be established within the USGCRP. The language in S. 1733 is terse and broadly defined, while H.R. 2454 includes much more detail, authorizes several subsidiary programs, and specifies the organizations to manage them. The bills differ on the design and location of the National Climate Service (NCS) office. S. 1733 places it within the Department of Commerce's National Oceanic and Atmospheric Administration (NOAA), while H.R. 2454 would establish a new interagency entity called the National Climate Service, plus a NOAA Climate Services Office, but leaves the evaluation of options, and design and location of the National Climate Service program, to the President. In H.R. 2454, the ultimate relationship of the NOAA Climate Services Office to the National Climate Service is left to be determined. The implementation plan of the NCS would be coordinated by the Director of OSTP.

Key recipients of climate services in both bills would be states, localities, and tribal governments, as well as the public, to enable the development and implementation of adaptation strategies to reduce vulnerability to climate variability and change. Stakeholders might need training in how to use more extensive climate information, as well as in dealing effectively with the wide uncertainty that is likely to continue to surround projections of climate, especially at more refined temporal and spatial scales.

State Adaptation Programs

Many policies and programs that influence the impacts of climate change on people, businesses, and natural resources are under the primary authority of states, who may in turn delegate authorities to local governments and may to some extent coordinate intergovernmental authorities at the local level. Both H.R. 2454 and S. 1733 appear to leave most authority and responsibility for addressing climate change adaptation to the states. Both bills do provide a role for the Environmental Protection Agency (EPA), which will facilitate the process of reviewing and approving state plans and disseminating "lessons learned" across states and tribes.

Both bills would require and help fund state and tribal adaptation programs. The Senate bill would require states to use the proceeds from the sales of allocated emission allowances exclusively for listed activities and as included in approved state climate change response plans, while the House bill identifies uses but provides a broader range of allowable activities that can be in compliance with approved state and tribal adaptation plans. Both bills use detailed and substantially identical language specifying the roles that states and tribes must play in developing and carrying out climate change adaptation plans.[22] S. 1733 and H.R. 2454 would provide financial resources to the states and tribes to support their adaptation planning, strategies, and implementation of certain measures through emission allowance allocations.

Public Health

Potential public health impacts of climate change include a wide range of risks including:

- a decline in air quality and an increase in allergenic pollen;
- more extreme temperatures;

- more frequent wildfires;
- altered conditions that foster the spread of communicable diseases and vector-borne diseases;
- events that threaten basic life support systems, such as droughts and floods which could adversely affect water, sanitation, and food systems.

The public health provisions in S. 1733 are essentially identical to those in H.R. 2454. Both bills mandate measures to assist health professionals in adapting to the health effects of climate change, including "the development, implementation, and support of State, regional, tribal and local preparedness, communication, and response plans to anticipate and reduce the threats of climate change." Both bills would require the Secretary of Health and Human Services (HHS) to develop a national strategy for public health adaptation, based on regular needs assessments and with input from an advisory board, to be updated every four years. Both bills would establish a Climate Change Health Protection and Promotion Fund.

The primary difference between the two bills is related to program funding and the use of the emission allowance allocations. S. 1733 would make funds obtained through revenue generated by emission allocation allowances available to the Secretary of HHS "without further appropriation," while H.R. 2454 makes funds available to the Secretary of HHS subject to further appropriation.

Natural Resources Adaptation

Adaptation of natural resources to climate change is a difficult concept to define, and determining a strategy to support it is complicated. To some extent, adaptation is occurring already: Trees in Alaska now grow much farther north than they did only 30 years ago. Small rodents in the Rockies are found at higher elevations than in the past. Freshwater marshes are being supplanted by salt-tolerant species. Fires are removing trees that cannot tolerate repeated droughts, and beetles that can now complete two generations in a year are speeding destruction of timber. Ecosystems lose species that are no longer able to find suitable habitat, or are unable to move rapidly enough to find it. Many of these changes are not considered desirable, and some constitute serious threats from an economic, public health, or aesthetic standpoint. Approaches by many parties have focused on lessening the negative effects of climate change, but more recently, planners have begun to plant different trees, map coastal areas to determine changes in tides and resulting vegetation, consider likely future ranges of species, and examine whether current use land patterns will permit species to reach more suitable habitats.

Both the House-passed and Senate-reported bills attempt to encourage federal and state strategies that can be adopted to support the resilience of species and the ecosystems on which they depend in the face of relatively rapid climate change. Within the provisions relating to natural resources, there are only a few major differences between the two bills. Both bills provide for federal, state, tribal, and local programs. Considerable emphasis at both levels is on planning, and within plans, on a tremendous number of factors to be developed, researched, or evaluated. In addition, both bills would create or facilitate the dissemination of new information sources. Both create a new National Climate Service, for example.

The natural resource adaptation activities in both bills would be funded through emission allowances, either through direct allocations or through revenues generated via auctioned allowances. The Senate bill would include a provision for land acquisition under the Land and Water Conservation Fund (LWCF), with funding available "without further appropriations," as has been proposed several times in the past. Such LWCF proposals have had considerable support from the scientific and environmental communities to protect rare ecosystems and/or recreational opportunities. However, opponents of LWCF—especially of LWCF proposals not subjected to annual appropriations oversight—have argued that the supervision of the appropriations process is necessary to protect property rights and landowners.[23] In addition, those that seek to limit federal spending in general may argue against allocation of money to any of the funds in S. 1733 in the absence of annual control by the appropriations and budget committees. H.R. 2454, on the other hand, requires that funds for natural resource adaptation programs be made available subject to annual appropriations.

In addition, both bills mandate the creation of new programs to disseminate data via geospatial information systems (GIS). While some data concerning wildlife already exist in GIS databases around the country, and considerable cooperation already exists among many agencies and academia, federal agencies and other levels of government might find the data useful for a variety of additional purposes, such as finding suitable locations for energy development and transportation infrastructure. Moreover, many see further coordination of geospatial data as essential for interagency and cross-sector coordination and planning.[24]

In both the House and Senate bills, there is little consideration of soils outside of carbon sequestration, biofuels, and alternative energy in relation to adaptation. Agricultural adaptation is also essentially absent from both bills. For example, at least six agencies within DOI are mentioned specifically, as is NOAA within the Department of Commerce, but within USDA, only the Forest Service figures prominently. USDA's Natural Resources Conservation Service (NRCS) is not mentioned in either bill, even though the agency's major responsibilities involve preventing soil erosion, protecting watersheds, and cooperating at multiple levels of government to control runoff and ease the effects of drought.

In addition, S. 1733 would address climate-change-exacerbated wildfire threats in several ways. It would define fire-ready communities, authorize cost-share grants to such communities, and direct cost-share agreements to encourage states and communities to become fire-ready. It also would direct mapping of fire risk in priority areas for fuel reduction treatments. Wildfires are not covered in the House-passed bill.

Water-Related Adaptation

Climate change is anticipated to affect water availability and use regionally, and may alter the frequency or intensity of water-related hazards, such as droughts and floods.[25] Potential impacts of climate change are of great interest to utility officials, federal agencies, and others concerned with water management and use. According to the Intergovernmental Panel on Climate Change (IPCC), higher water temperatures, increased precipitation intensity, and longer periods of low flows will exacerbate many forms of water pollution, with impacts on water system reliability and operating costs, human health, and ecosystems. Similarly, climate change affects the function and operation of existing water infrastructure, as well as water management practices.[26] Temperature change drives other changes in natural environmental processes that, in turn, affect the quality and quantity of water resources. A range of impacts are anticipated, although they are likely to vary by region, including warmer water, precipitation changes, loss of reservoir storage and snowpack, sea level rise, increases in storm intensity, increased risk of flood damage, water treatment and distribution challenges, increased wastewater treatment needs and costs due to heavier runoff, and increased demand in response to heat waves and dry spells.

Water-related adaptation is likely to incorporate a range of measures: demand management and conservation, difficult land use choices in at-risk areas, investments in infrastructure, and aquatic ecosystem protection and restoration. Adapting to climate's water-related effects presents significant challenges, in part because of the wide variety of entities involved in managing and using water,[27] and the ecosystems and species that depend on its availability, variability, and quality. Responsibilities for, and funding of, different water adaptation measures are at issue as Congress considers climate change legislation.

S. 1733 would include provisions specific to water-related adaptation; no similar provisions are included in H.R. 2454.[28] To assist adaptation by water utilities, S. 1733 (in Division A) would establish a research program to assist drinking water utilities (Section 211) and a program of grants to states and Indian tribes for water system adaptation projects (Section 381). S. 1733 would also include two other water-specific adaptation provisions—Section 382, which would establish a grants program to states and Indian tribes for adapting to climate-related flood impacts, and Section 384, which would provide assistance to coastal (including Great Lakes) states for adapting to climate change. No specific provisions were included for drought or for adaptation of agricultural or energy sector water use to changed water resource availability and quality. Sections 381, 382, and 384 would be funded through the state climate change response account (Section 210 of Division B); Section 211 of Division A includes only an authorization of appropriations. These provisions focus to a greater extent on adapting to water *quantity* challenges of climate change (e.g., ensuring utilities' reliable delivery of water supply) than water *quality* (e.g., changes in dissolved oxygen levels or water chemistry resulting from warmer temperatures).

Water-related adaptation planning and measures also are covered under broader provisions, most notably federal assistance for state adaptation efforts, federal research and assessment, and natural resources adaptation efforts by designated federal agencies. The bills would provide funding for select federal agencies to undertake aquatic ecosystem restoration activities and other water-related natural resource adaptation actions using allocations established for natural resources adaptation;[29] otherwise, they do not specifically allocate funding for federal agencies to undertake climate change adaptation for water infrastructure (e.g., dams, levees, navigation improvements) or other water-related programs under their jurisdiction.[30] The bills would direct federal water resource agencies, including the Bureau of Reclamation and the Army Corps of Engineers, to adapt their plans, programs, and activities. Because most water resource projects typically receive project-specific authorization from Congress, it is unclear how much authority and funding these agencies would have to implement adaptation actions.

S. 1733 and H.R. 2454 largely focus their federal natural resources adaptation provisions on the agencies responsible for managing and protecting water resources, such as the U.S. Army Corps of Engineers, the Bureau of Reclamation, and the U.S. Environmental Protection Agency, not the agencies working with the users that depend on water resources. For example, neither U.S. Department of Agriculture agencies (e.g., Natural Resources Conservation Service) nor Department of Energy entities (e.g., power marketing administrations) are included. Agriculture, particularly in the West, is the largest consumer of water. The energy sector also withdraws significant quantities during extraction, processing, and generation. Water-related independent entities, like the Tennessee Valley Authority, also are not specifically addressed in the natural resources adaptation provisions. One exception is that S. 1733 would include the Federal Emergency Management Agency (FEMA) in the natural resources adaptation panel proposed in the bill (Section 365). FEMA manages flood hazard mitigation programs and the national flood insurance program; FEMA is not included in a similar provision of H.R. 2454 (Section 475).

In summary, the water-specific provisions of S. 1733, which have no comparable provisions in H.R. 2454, would be focused largely on issues arising from water quantity changes (e.g., reduced municipal water supplies, increased flooding, higher sea levels). While broad natural resource provisions of the bills include water resource agencies and funding for aquatic ecosystem restoration, less attention is given in S. 1733 or H.R. 2454 to the adaptation of federal water resources infrastructure to changes in water resource quantity and quality; similarly, while S. 1733 would address some of the adaptation challenges faced by municipal water providers, little attention is given in either bill to managing adaptation in two of the largest water use sectors—agriculture and energy.

Appendix. Comparison of Adaptation-Related Provisions in H.R. 2454 (as Passed by the House) and S. 1733 (as Reported by the Senate EPW Committee)

S. 1733 (as reported by the Senate EPW Committee)	H.R. 2454 (as passed by the House)	Comments
International Climate Change Adaptation		
Sec. 324. International Climate Change Adaptation and Global Security Program. The Secretary of State, consulting with heads of other agencies, is required to establish an International Climate Change Adaptation and Global Security Program. After consulting with heads of agencies, the Secretary of State or other Presidential designee directs the distribution of funding to assist vulnerable countries and their populations within them, and for programs that promote U.S. interests by supporting adaptation to climate change. Funding may be provided bilaterally, and/or through multilateral or international institutions under the United Nations Framework Convention on Climate Change (UNFCCC).	**Sec. 493. International Climate Change Adaptation Program.** The Secretary of State, consulting with heads of other agencies, is required to establish an International Climate Change Adaptation Program. Assistance must supplement, not supplant, other resources for similar activities. After consulting with heads of agencies, the Secretary of State or other Presidential designee directs the distribution of allowances to assist vulnerable countries and populations within them, by supporting adaptation to climate change. Funding may be provided bilaterally, and/or through multilateral or international institutions under the United Nations Framework Convention on Climate Change (UNFCCC). **Sec. 494.** Multilateral or international recipients must receive 40% to 60% of distributions, and must meet eligibility and reporting requirements, overseen by the Secretary of State. **Sec. 495. Bilateral Assistance.** USAID may carry out programs and give allowances to any private or public group to assist with the development of adaptation plans and projects to assist the most vulnerable developing countries, support investments, research programs and activities, and encourage engagement of local communities. No more than 10% of the allowances distributed for	Language in the two bills is nearly identical on establishment of the program and distribution of allowances/assistance. S. 1733 provides for "additionality" of resources as demanded by international guidance for accounting for financial commitments under the Climate Convention. On uses of assistance, the two bills are similar, but S. 1733 omits much of the detail and prescriptive language contained in H.R. 2454. Specific differences include: • S. 1733 distributes funds while H.R. 2454 distributes allowances; • S. 1733 sets no limits on the portion of assistance to distribute multilaterally, while H.R. 2454 requires 40-60% go to multilateral funds or mechanisms. • H.R. 2454 sets eligibility criteria for multilateral funds or institutions to receive allowances • H.R. 2454 authorizes USAID to provide assistance with specified purposes, limited to no more than 10% to any single country in a year. • H.R. 2454 prescribes priorities and conditions for USAID's use of assistance. Content of the reports and reviews also differs. S. 1733 allows discretion in bilateral programs to involve any agency. H.R. 2454 does not explicitly

	bilateral assistance in a year may support activities in any one country. The USAID Administrator must provide for consultation and disclosure of information to stakeholders regarding any programs or activities carried out under this section.	provide for bilateral programs in agencies other than USAID, though other agencies may have existing or potential expertise and programs that could support international capacity-building, technological, financing or other needs related to adapting to climate change.
Sec. 325. Evaluation and Reports. Directs the Board to establish a system to monitor and evaluate the international climate change assistance. Reports to Congress required within one year of enactment, within three years of enactment, and then triennially, to review needs and opportunities for further investment in developing countries.	**Sec. 495(d). Annual Reports.** The USAID Administrator must report to the President and to Congress within 180 days after enactment and annually thereafter on the following: the extent of adverse climate change impacts in the most vulnerable developing countries; the potentially destabilizing effects of climate change affecting U.S. national security; how emission allowances were distributed and recommendations for future years; and the status of international cooperation. **Sec. 495(e). Monitoring and Evaluation.** The Administrator of USAID must establish performance goals, indicators, and other means to evaluate, *inter alia*, the degree to which local communities were informed of and engaged in, activities; the impacts of adaptation activities; and recommendations for adjustments.	Both bills require reviews and evaluations of international assistance for adaptation, as well as reports to Congress. H.R. 2454 is more detailed and prescriptive regarding how to monitor and evaluate the programs, and content of the reports.
Sec. 207. International Climate Change Adaptation and Global Security. Directs allocation of allowances to climate change adaptation. The quantity of allocation is specified in Sec. 771(a)(14): 2012-2021: 1.0% of annual allowances; 2022-2026: 2.0% of annual allowances; 2027-2050: 5.0% of allowances.	**Sec. 782(n) International Adaptation.** Directs the EPA Administrator to allocate emission allowances for international adaptation for: 2012-2021: 1.0% of annual allowances; 2022-2026: 2.0% of annual allowances; 2027–2050: 4.0% of allowances.	Almost identical allocations of allowances, except for the latest period of 2027-2050, with a 5% distribution under S. 1733 versus the 4% under H.R. 2454. The rising percentages of allowances may correspond with predicted increases in climate change, and associated accelerating adaptation needs.

Domestic Climate Change Adaptation		
Sec. 341. National Climate Change Adaptation Program. The President must establish a National Climate Change Adaptation Program within the United States Global Research Program (USGCRP) to increase effectiveness of federal adaptation efforts.	No similar provision.	H.R 2454 and S.1733 take different approaches to adaptation to climate change at the national scale. S 1733 directs the President to establish a National Climate Change Adaptation Program, leaving broad discretion as to the new program's organization, strategy, contents, etc. In contrast, H.R. 2454 primarily expands the existing U.S. Global Change Research Program more explicitly to emphasize the *effects* of climate change, and to add impact and adaptation-related research, new observational, research, and to improve information based decision-making efforts (See Sec. 451 below).
No similar provision	Sec. 451.Global Change Research and Data Management. Repeals and replaces Titles I and III of the existing Global Change Research Act (GCRA) of 1990 (P.L. 101-606; 15 U.S.C.2921 et seq.). Directs the President to establish an interagency coordinating committee, a U.S. Global Change Research Program (USGCRP), a National Global Change Research and Development Plan, budget coordination, Vulnerability Assessments, Policy Assessments, and annual reports to Congress. It also establishes a Global Change Research Information Exchange and interagency data management, and requires reports on ice sheet melt and sea level rise, and on implications of hurricane frequency and intensity patterns. Establishes the Office of Science and Technology Policy as the "lead agency" and authorizes $10 million annually for FY2009-FY2014 for "interagency program activities." In Sec. 451(5), the National Global Change Research and Assessment Plan must, inter alia,	The GCRA of 1990 established an interagency coordinating committee and the U.S. Global Change Research Program (USGCRP). H.R. 2454 provisions are in many aspects similar or identical to those in the GCRA, but more expansive. It leaves in place Title II of the GCRA, which covers international global change research cooperation. Compared to the existing management of the USGCRP, H.R. 2454 would make the White House Office of Science and Technology the lead agency. To the 1990 purpose is added "observation" and "outreach" activities, with an emphasis on "effects" of global change. The Global Change Research Program is (re)established in para. (4) "to respond to the information needs of communities and decision-makers and to provide periodic assessment of the vulnerability of the United States and other regions." Other provisions, however, do not maintain this more expansive language, confining the provisions to "research" (e.g., for interagency coordination). In congruity, the bill would expand

	catalog types of information needed by decision makers to develop policies to reduce vulnerabilities to global change, and provide for economic, demographic, technological, and other information to meet the needs of decision-makers. Sec. 451(8) requires a Policy Assessment within one year of enactment and every four years thereafter by the National Academy of Public Administration and the National Academy of Sciences, to cover both climate change mitigation and adaptation options.	participation in the interagency coordinating committee to include representation not just of science programs but also resource management and climate mitigation agencies and programs. Unlike the existing program, the Office of Science and Technology Policy is made the "lead agency." OSTP does not have existing authority to "allocate funds" to agencies. In H.R. 2454, the relationship of the Global Change Research Information Exchange to the new National Climate Services (Sec. 452) and the new Climate Service Office in NOAA (Sec. 452(e)) is not defined, and is to be established or designated by the President.
Sec. 157. Study of Risk-Based Programs Addressing Vulnerable Areas. The Administrator of the EPA, or other presidentially- designated heads of federal agencies, must conduct a study that reviews and assesses federal pre-disaster mitigation, emergency response, and flood insurance policies and programs affecting areas vulnerable to climate change; describe better strategies to address vulnerabilities, and whether existing federal policies support state response and adaptation goals in Sec. 211. The studies also must identify and recommend how to resolve contradicting programs that address areas vulnerable to climate change, and identify annual cost savings that could be achieved with recommended strategies. Report is due to Congress within two years of enactment	Sec. 451(7). Requires a Vulnerability Assessment within one year of enactment and every five years thereafter, with a time frame of the subsequent 25 to 100 years. Assessment is to cover the United States and other world regions, and multiple sectors and categories of impacts.	While H.R. 2454 requires comprehensive and periodic assessments of vulnerability to climate change, S. 1733 authorizes one report, confined to specific types of disaster (not necessarily due to climate change). The S. 1733 study is a one-time requirement aimed at evaluating expected cost savings from improving inter-agency and inter-governmental coordination rather than a broad vulnerability assessment as authorized in H.R. 2454.

Sec. 342. Climate Services. The Secretary of Commerce acting through the Administrator of the	Sec. 452. National Climate Service. Establishes a National Climate Service (NCS), and defines the activities to be	Both bills would establish a new Climate Service program. S. 1733 places the National Climate Service within NOAA.
S. 1733 (as reported by the Senate EPW Committee)	**H.R. 2454 (as passed by the House)**	**Comments**
National Oceanic and Atmospheric Administration (NOAA) must establish a National Climate Service within NOAA. The National Climate Service is to: • develop climate information, data, forecasts and warnings at national and regional scales; and • distribute information related to climate impacts to state, local, and tribal governments and the public to help develop and implement strategies to reduce vulnerabilities to climate variability and change.	undertaken within the National Oceanic and Atmospheric Administration (NOAA), to: • advance understanding of climate variability and change at different scales; • provide forecasts, warnings, and other information on weather and climate. Its goal is to meet the needs of decisionmakers in federal agencies; state, local, and tribal governments; regional entities; and other stakeholders and users, for information related to climate variability and change. Requires a report to Congress within two years of enactment to describe institutions and propose how to establish a National Climate Service. Requires the Undersecretary of NOAA to establish a Climate Services Office within NOAA, and to establish a Clearinghouse of Federal Climate Service Products and Links to Federal Agencies Providing Climate Services. Requires a number of additional programs and services to support climate change information and adaptation planning. Sec. 452(m) specifies that nothing in Sec. 452 authorizes requirements for states, tribes or local governments to develop adaptation or response plans or to take any other actions in response to variations in climate that may impose a financial burden to such governments.	H.R. 2454 leaves evaluation of options, and design and location of the national program to the President while also establishing an office within NOAA. In H.R. 2454, the relationship of the National Climate Service, or the new Climate Service Office in NOAA, to the Global Change Research Program and the Global Change Research Information Exchange is not made explicit. "Climate" and "climate variability," as distinct from "weather," are not defined, and have been inconsistently used in some proposals for national climate services. Sec. 452(b)(2) explicitly calls for "weather" forecasts, warnings and other information.

State and Tribal Programs			
Sec. 210. State Programs for Greenhouse Gas Reduction and Climate Adaptation. Within 2 years of enactment, the EPA Administrator or other presidential designee(s) must promulgate regulations to implement this section. Of each vintage year's allowances specified in Sec. 771(a) for state adaptation, the EPA Administrator must reserve: • 10% for coastal and Great Lake States, for purposes in Sec. 384 (see below); • 10% for states for wildfire programs for purposes in Sec. 383 (see below);	Sec. 453. State Programs to Build Resilience to Climate Change Impacts. Sec. 453(b) Within two years of enactment, the EPA Administrator or other presidential designee(s) must promulgate regulations to implement this section. From 2011-2049, the EPA Administrator or other federal agency head(s) designated by the President must distribute allowances to states and tribes annually. States receive allowances on the basis of (1) population and (2) the ratio of each state's per capita income relative to that of the United States as a whole. Tribes receive 1% of allowances, distributed competitively based on their adaptation plan or project proposals.	Requirement to promulgate implementing regulations is identical. S. 1733 directs the EPA Administrator to reserve percentages of allowances allocated to state adaptation programs for specific states and uses, as well as tribes. The remainder is distributed to states and tribes by formulae. H.R. 2454 does not include these "reserve" paragraphs, beginning directly with distributions of allowances. The formulae and methods for determining each State's allowances are identical in H.R. 2454 and S. 1733, although the language varies slightly. In S. 1733, the initial combined State Climate Change Response and Transportation Fund in Treasury has been	
S. 1733 (as reported by the Senate EPW Committee)	**H.R. 2454 (as passed by the House)**	**Comments**	
• at least 1% for Indian tribes, of which at least 18% must go to Alaska Native Villages; and distribute the remainder of allowances for State government programs for GHG reduction and climate adaptation. Allowances or proceeds from auction of allowances are deposited into State Climate Change Response (SCCR) accounts. From 2011-2049, the EPA Administrator or other federal agency head(s) designated by the President must distribute allowances for the subsequent calendar year to states and tribes annually. States receive allowances generally on the basis of (1) population and (2) the ratio of each State's per capita income relative to that of the United States as a whole. States must distribute at least 12.5% of the	Tribes with adaptation plans have priority in distribution. Uses of allowances are listed, with priority being given to reduce flood risks. Allowances must be sold within one year, with proceeds deposited into the State Energy and Environment Development (SEED) accounts. States the intention of Congress that funds provided should supplement, not replace, existing sources of funding.	eliminated and replaced with separate funds for transportation and state climate change responses.	

| proceeds deposited to SCCR accounts to local governments to address specific adverse impacts of climate change (listed below). States and tribes shall use the allowance proceeds exclusively to develop and implement policies, programs or measures that reduce GHG emissions or build resilience to climate change via activities listed under Sec. 221(g)(2). Funds must be used in accordance with approved state or tribe climate change response plans, and only for specific activities, to address:
• water system partnerships (Sec. 381);
• flood control, protection, prevention and response programs (Sec. 382);
• impacts on water quality, supply or reliability of state-owned or operated water systems (Sec. 381(d));
• recycling (Sec. 154);
• adverse climate change impacts on agricultural or ranching activities;
• projects to restore abandoned mine lands that increase carbon sequestration or reduce GHG emissions while providing other benefits;
• adverse impacts on air pollution or air quality;
• measures to reduce GHG emission that decrease other air pollutant emissions as well.
At least 12.5% of allowance proceeds in SCCR accounts must be distributed to local governments for activities listed under (2) above.
States and localities shall ensure that funds are | |

	used to assist categories of "socially and economically vulnerable populations." • States the intention of Congress that funds provided should supplement, not replace, existing sources of funding.	
Sec. 210(h). State and Tribal Response Plans. In order to receive funds, states and tribes must have approved adaptation plans. Beginning with vintage year 2012, states must have approved State climate change response plans to meet regulations to be promulgated under Sec. 221(b), with elaboration of content under Sec. 221(g). The state climate change response plans must, at a minimum, assess and prioritize vulnera-bilities; identify and prioritize cost-effective projects, programs, and measures to mitigate and build resilience to current and predicted climate; assess potential carbon reductions by changing land management policies; ensure that the state consider and undertakes a variety of listed types of initiatives; consider impacts on socially and economically vulnerable populations; use pre-disaster mitigation, emergency response, and public insurance programs; and be consistent with federal conservation and environmental laws and try to avoid environmental degradation. Plans must be revised and resubmitted every five years. Tribal climate change response plans have same requirements as the states, but may vary if necessary to account for special circumstances of Indian tribes.	**Sec. 453. State Programs to Build Resilience to Climate Change Impacts.** In order to receive funds, states and tribes must have approved adaptation plans. Beginning with vintage year 2015, states and tribes must have approved State climate change response plans to meet regulations to be promulgated under Sec. 453(b), with elaboration of content under Sec. 453(f). State and Tribal climate change response plans must, at a minimum, assess and prioritize vulnerabilities; identify and prioritize cost-effective projects, programs, and measures to mitigate and build resilience to current and predicted climate; assess potential carbon reductions by changing land management policies; ensure that the state considers and undertakes a variety of listed types of initiatives; and be consistent with federal conservation and environmental laws and try to avoid environmental degradation. Plans must be revised and resubmitted every five years.	Language in the two bills is similar, except: Allocation of allowances under this section are contingent on approved state adaptation plans by 2012 under S. 1733 and by 2015 under H.R. 2454. S. 1733 additionally specifies that states must consider and undertake a longer list of requirements, where appropriate, protect forested land using science-based ecological restoration practices, and consider impacts on socially and economically vulnerable populations. S. 1733 allows adaptation funds to be used for carbon sequestration on abandoned mine lands. Reporting and enforcement language is identical in both bills. In addition, S. 1733 has an auditing provision that gives authority to the EPA Administrator or other presidential designee, to audit or review implementation and compliance of state plans. No auditing provision exists in H.R. 2454. In both bills, the EPA Administrator must take into account lessons learned, avoid duplication, and coordinate with state natural resources adaptation plans.

Public Health		
Sec. 353. National Strategic Action Plan. Requires the Secretary of Health and Human Services (HHS) to prepare a national strategic action plan to prepare for and respond to public health impacts of climate change in the United States and other nations, in consultation with relevant agencies and stakeholders. The plan must be revised by 2014 and every four years thereafter. Requires a public health needs assessment from the National Research Council and the Institute of Medicine within one year of enactment.	**Sec. 463. National Strategic Action Plan.** Similar to Senate bill except gives authority to conduct and fund research to the Secretary of HHS, directed by the Director of the Centers for Disease Control and Prevention, and the head of any other appropriate federal agency.	
Sec. 354. Advisory board. Establishes an advisory board to provide scientific and technical advice to the Secretary of Health and Human Services on domestic and international impacts of climate change on human health.	**Sec. 464. Advisory Board.** Essentially identical to Senate bill.	
Sec. 355. Reports. Describes the requirement for reports on a needs assessment, due within one year of enactment, and on climate change health protection and promotion, due by July 1, 2013 and every 4 years thereafter.	**Sec. 465. Reports.** Essentially identical to Senate bill.	
Sec. 356. Definitions. Provided definitions for health impact assessment, national strategic action plan, and secretary.	**Sec. 466. Definitions.** Essentially identical to Senate bill.	
Sec. 211. Climate Change Health Protection and Promotion Fund. Establishes in the Treasury a Climate Change Health Protection Fund which will receive revenue from the auctioning of 0.1% of each year's emission allowances. The funds are available "without further appropriation" and should supplement existing sources of funding. The Secretary of HHS may distribute funds from	**Sec. 782(l)(2). Domestic Adaptation.** Directs the EPA Administrator to allocate 0.1% of emission allowances for the Climate Change Health Protection and Promotion Fund (Sec. 467) in 2012 and thereafter. Availability of funds would be subject to further appropriation.	Language in the two bills is similar except: Funds are available to Secretary of HHS subject to further appropriation in H.R. 2454, while S. 1733 makes funds available "without further appropriation" or fiscal year limitation.

the Fund to federal agencies, other governments, or other entities, to carry out any of the provisions of the health and climate change provisions in this subtitle.		
Natural Resources Adaptation		
Sec. 361. Purposes. Purposes of this subpart are to establish an integrated program that responds to climate change, including ocean acidification, drought, flood-ing, and wildfire, and to provide financial support and incentives for these activities.	**Sec. 471. Purposes.** Similar to Senate bill.	Senate bill makes specific mention of drought, flooding, and wildfire; House bill does not.
S. 1733 (as reported by the Senate EPW Committee)	**H.R. 2454 (as passed by the House)**	**Comments**
Sec. 362. Natural Resources Climate Change Adaptation Policy. States that federal policy is "to use all practicable means to protect, restore, and conserve natural resources so that natural resources become more resilient, adapt to, and withstand the ongoing and expected impacts of climate change, including, where applicable, ocean acidification, drought, flooding, and wildfire."	**Sec. 472. Natural Resources Climate Change Adaptation Policy.** States that federal policy is "to use all practicable means and measures to protect, restore, and conserve natural resources to enable them to become more resilient, adapt to, and withstand the impacts of climate change and ocean acidification."	Essentially the same except the Senate bill specifically mentions drought, flooding, and wildfire, while the House bill does not.
Sec. 363. Definitions. Defines 15 terms used in the subpart: account, administrators, board, center, coastal state, corridors, ecological processes, habitat, Indian tribe, natural resources, natural resources adaptation, panel resilience/resilient, state, and strategy.	**Sec. 473. Definitions.** Defines nine terms used in the subpart,: coastal state, corridors, ecological processes, habitat, Indian tribe, natural resources, natural resources adaptation, resilience/resilient, and state.	Neither bill includes the consideration of air and soil resources in the definition of natural resources. In addition, House bill definition of "natural resources" mentions land and water while the Senate bill omits these terms. In the definition of "ecological processes, both bills contain the phrase "biological, chemical, or physical interaction," but it should be noted that these processes, are not mutually exclusive.

Sec. 364. Council on Environmental Quality. Directs Chair of the Council on Environmental Quality (CEQ) to advise the President on developing and implementing a Natural Resources Climate Change Adaptation Strategy and federal natural resource agency adaptation plans, and to coordinate such activities.	**Sec. 474. Council on Environmental Quality.** Essentially identical to Senate bill.	
Sec. 365. Natural Resources Climate Change Adaptation Panel. Establishes a Natural Resources Climate Change Adaptation Panel as a forum for coordinating development and implementation of the federal adaptation strategy. The Chairperson of CEQ is to chair the Panel. The Panel must be established within 90 days of enactment of the law, and include NOAA, USFS, NPS, FWS, BLM, USGS, Reclamation, BIA, EPA, Army COE, CEQ, FEMA, and other federal agencies with jurisdiction over natural resources, as determined by the President.	**Sec. 475. Natural Resources Climate Change Adaptation Panel.** Similar to Senate bill.	The only difference is that the Senate bill includes the Federal Emergency Management Agency (FEMA) on the Adaptation Panel, while the House bill does not. Neither bill includes the USDA's Natural Resources Conservation Service specifically on the Adaptation Panel, although President has discretion to add other agencies.
Sec. 366. Natural Resources Climate Change Adaptation Strategy. Describes the climate change adaptation strategy to be developed by the panel established in Sec. 365. The strategy must be developed within one year of enactment of the subpart, and must be reviewed and revised every five years. The strategy must be based on the best available science; must be developed in cooperation with states, Indian tribes, other federal agencies, local governments, conservation organizations, scientists, and other stakeholders; and must be open for public comment. The purpose of the strategy is to	**Sec. 476. Natural Resources Climate Change Adaptation Strategy.** Similar to Senate bill.	Differences are minor and include multiple references to "ongoing" and "expected" or "expanding" impacts in the Senate bill, while the House bill regularly refers to "ocean acidification" in concert with climate change.

protect, restore, and conserve natural resources to enable them to become more resilient, adapt to, and withstand the impacts of climate change, and to identify opportunities to mitigate ongoing and expected impacts.		
Sec. 367. Natural Resources mogram, to be led by the USGS National Climate Change and Wildlife Center (established by this section) and the National Climate Service in NOAA. Program is to provide technical assistance, conduct and sponsor research, and provide research, monitoring tools, and information. Secretaries of Commerce and the Interior must conduct initial and then five-year surveys of natural resources impacts of climate change; monitoring of baselines and trends; and stakeholder needs for monitoring, research, and decision tools. Establishes a Science Advisory Board to advise Secretaries on impacts and scientific strategies and mechanisms, and to identify and recommend research priorities.	**Sec. 477. Natural Resource Adaptation Science and Information.** Similar to the Senate bill.	Minor differences include multiple references to "ongoing" and "expected" or "expanding" impacts in the Senate bill, while the House bill regularly refers to "ocean acidification" in concert with climate change, where the Senate bill specifically also mentions drought, flooding, and wildfire.
Sec. 368. Federal Natural Resource Agency Adaptation Plans. Requires each federal agency represented on the Natural Resources Climate Change Adaptation Panel to complete a Natural Resources Climate Change Adaptation Plan, consistent with the policy under Sec. 472, within one year of enactment. After approval by the President, adaptation plans must be submitted to specified congressional committees (e.g., House Natural Resources; Senate Energy and Natural Resources; Environment and Public Works; and any others with agency jurisdiction) within 30 days of approval.	**Sec. 478. Federal natural resource agency adaptation plans.** Similar to the Senate bill, though Senate Environment and Public Works is not mentioned specifically.	Senate bill requires the agency action plans to include "any changes in decisionmaking processes necessary to increase the ability of resources under the jurisdiction of the department or agency and, to the maximum extent practicable, resources under the jurisdiction of other departments and agencies that may be significantly affected by decision of the department or agency, to become more resilient, adapt to, and withstand the ongoing and expects impacts of climate change....." House bill lacks similar provision.

S. 1733 (as reported by the Senate EPW Committee)	H.R. 2454 (as passed by the House)	Comments
Sec. 369. State natural resources adaptation plans. Requires states to prepare a state natural resources climate change adaptation plan to be eligible to receive funds under Sec. 370. The plan must include priorities, programs, and measures of effectiveness, and must be reviewed and updated every five years.	**Sec. 479. State natural resources adaptation plans.** Similar to the Senate bill.	House and Senate bills are essentially identical, except Senate bill adds a few additional items to include in plans. Note that there is a typographical error in Sec. 369(e)(5): where "regional fishery management plants" should read "regional fishery management plans."
Sec. 370. Natural Resources Climate Change Adaptation Account. Overall, section distributes allowances to states for adaptation activities, and distributes, "without further appropriation," proceeds from auction of allowances to specified federal agencies and programs. States must pay at least 10% of costs of any federal grant provided in this section. (See specific subsections, below.)	**Sec. 480. Natural Resources Climate Change Adaptation Fund.** Overall, section distributes allowances from Sec. 721(a) to support state adaptation activities, and funds from auction of allowances to support specified federal agencies and programs. Specifies that appropriation levels for both federal programs should be no less than the proceeds from specified allowances and auction of allowances. States must pay at least 10% of costs of any federal grant provided in this section. (See specific subsections, below.)	In these two sections, the bills are very similar in their details and structure, and their allocations to various programs generally differ by less than a percentage point. However, the major difference is that the Senate bill provides funds "without further appropriation," while the House bill subjects them to annual appropriations. The House provisions rely on the creation of special funds and provide the strongest possible encouragement to the appropriations committees to approve appropriations at the full authorized levels for the new funds.
Sec. 370(a)(1). Distributes allowances from Sec. 771(a)(16) and Sec. 216 (which allocates allowances from Sec. 771(a)(16) to a new Natural Resources Climate Change Adaptation Account, NRCCAA) for Wildlife Restoration Program (84%; see 16 U.S.C. 669c) and Coastal Zone Management Act (CZMA) (16%; see 16 U.S.C. 1455(c)).	**Sec. 480(a).** Directs a percentage of the emission allowances in Sec. 721(a) to state adaptation activities under Sec. 479 for wildlife restoration grants (84.4%), and coastal zone conservation (15.6%).	The House and Senate bills both allocate allowances to state programs, rather than proceeds from auction of allowances.
Sec. 370(a)(2). Distributes proceeds from auction of allowances under Sec. 771(b)(7) and Sec. 212. Allocates proceeds to the Department of	**Sec. 480. Natural Resources Climate Change Adaptation Fund.** Establishes a new Natural Resources Climate Change Fund (NRCCF) in Treasury, with appropriations authorized at not	In S. 1733, Sec. 370(a)(2)-(6) all receive funding from Sec. 771(b)(7) and Sec. 212. However, both Sec. 771(b)(7) and Sec. 212 direct that their proceeds go only to Sec. 370(a)(2). If these latter

	Comments
Interior (DOI) as follows: • 28%—specified natural resources adaptation activities by DOI agencies and Sec. 371 Corridors Information Program. • 8%—specified programs for endangered species, wetlands, migratory birds, coastal program, and private lands. • 5%—specified tribal programs under Bureau of Indian Affairs (BIA) and the Fish and Wildlife Service (FWS). less than estimated total annual deposits to Natural Resources Climate Change Adaptation Fund (NRCCAF). Sec. 480(c)(1) allocates funds from NRCCF to DOI in a manner similar to S. 1733, except amounts are 27.6%, 8.1%, and 4.9%, respectively.	sections are correct, then the source of funding for Sec. 370(a)(3)-(6) is not clear. It seems likely that the intent in Sec. 771(b)(7) and Sec. 212 was to include all five of these paragraphs as eligible for funding.
S. 1733 (as reported by the Senate EPW Committee)	
H.R. 2454 (as passed by the House)	
Sec. 370(a)(3). Directs 20% of funds available from Sec. 771(b)(7) and Sec. 212 under this subpart, for Land and Water Conservation Fund (LWCF)-type purposes—1/6 for Interior's stateside assistance; 1/3 for Interior land acquisition; 1/6 for Forest Service grants for land or easement acquisition; and 1/3 for Forest Service land purchases—with considerations for funding allocation. Amounts allocated to LWCF to be available "without further appropriation." **Sec. 480(c)(2).** Essentially identical to Senate bill, except 19.5%, rather than 20% for LWCF-type purposes. Amounts allocated to LWCF are subject to annual appropriations.	In the Senate bill, the availability of funds under the LWCF is available "without further appropriation." While a comparable proposal has had considerable support from the scientific and environmental communities to protect rare ecosystems and/or recreational opportunities, opponents have argued that the supervision of the appropriations process is necessary, in order to protect property rights and landowners. In addition, persons wishing to limit federal spending in general may argue against allocation of money to either fund in the absence of annual control by appropriations and budget committees. In S. 1733, also see comments under Sec. 370(a)(2) concerning funding from Sec. 771(b)(7) and Sec. 212.
Sec. 370(a)(4). Directs 8% of funds available from Sec. 771(b)(7) and Sec. 212 under this subpart for natural resource adaptation by the Forest Service on the national forests and national grasslands and through financial and technical assistance.	**Sec. 480(c)(3).** Essentially identical to Senate bill, except 8.1%, rather than 8%, for the same purposes.	In S. 1733, also see comments under Sec. 370(a)(2) concerning funding from Sec. 771(b)(7) and Sec. 212.

Sec. 370(a)(5). Directs 11% of funds available from Sec. 771(b)(7) and Sec. 212 to Secretary of Commerce for specified coastal, estuarine, fishery, marine mammal, endangered species, and coastal programs.	**Sec. 480(c)(4).** Essentially identical to Senate bill, except 11.5%, rather than 11%, for the same purposes.	In S. 1733, also see comments under Sec. 370(a)(2) concerning funding from Sec. 771(b)(7) and Sec. 212.
Sec. 370(a)(6). Directs 12% of funds available from Sec. 771(b)(7) and Sec. 212 to EPA and 8% to Corps of Engineers for specified estuarine and freshwater ecosystem protection programs, including programs in a list of 20 named ecosystems, as well as water resources programs.	**Sec. 480(c)(5).** Essentially identical to Senate bill, except 12.2% for EPA and 8.1% for the Corps of Engineers, rather than 12% and 8% respectively, for the same purposes. Slightly different list of named ecosystems.	In S. 1733, also see comments under Sec. 370(a)(2) concerning funding from Sec. 771(b)(7) and Sec. 212.
Sec. 371. National Fish and Wildlife Habitat and Corridors Information Program. Establishes a National Wildlife Habitat and Corridors Information Program within DOI to support states and tribes to develop coordinated geographic information system (GIS) of fish and wildlife habitat and corridors for information and modeling of climate change impacts and adaptation, and to enhance state and tribal wildlife action plans. Use of GIS intended to aid policy makers at all levels.	**Sec. 481. National Wildlife Habitat and Corridors Information Program.** Essentially identical to Senate bill.	Neither bill specifies a funding level, but funding available from Sec. 370(a)(2) in S. 1733, and from Sec.480(c) in H.R. 2454. GIS data bases for many areas of wildlife management already exist, though coverage is often spotty and comparisons may be difficult. Major benefit of programs could be increased utility from better coordination and compatibility.
Secretary authorized to support states and tribes financially and technically to develop and implement system.		
Sec. 372. Additional Provisions Regarding Indian Tribes. Specifies that nothing in this subpart amends federal trust responsibilities to Indian tribes. Exempts from Freedom of Information Act (FOIA) disclosure any information relating to sacred sites or cultural activities that tribes consider confidential. Clarifies that DOI Secretary may apply provisions of the Indian Self-Determination	**Sec. 482. Additional Provisions Regarding Indian Tribes.** Contains similar provisions. FOIA exemption is more detailed and provides that information received by a federal agency concerning human remains, resources, cultural items, activities identified by an Indian tribe as traditional or cultural, is protected from FOIA disclosure if head of agency, in consultation with DOI Secretary and tribe, determines that	Both sections specify that the DOI Secretary may authorize an Indian tribe to implement DOI climate change activities related to natural resources conservation in this subpart.

and Education Assistance Act in implementing this subpart regarding safeguards for natural resources conservation. Protects rights reserved under treaties for tribes to take certain plant foods.	disclosures may cause significant invasion of privacy, risk harm to remains or items, or impede site use.	
Sec. 383. Wildfire. Defines fire-ready communities and authorizes cost-share grants to such communities. Directs federal fire agreements to encourage communities to become fire-ready. Directs fire risk mapping of priority areas needing fuel reduction efforts.	No similar provision	S. 1733 authorizes a program to reduce the risk of wildfires in fire-ready communities and establishes criteria therein. The program creates regional maps of communities most at risk of wildfire and identifies priority areas and identifies several examples for priority areas needing "hazardous fuel treatment and maintenance." Grants would be provided for fire protection education programs, training programs for local firefighters, equipment to increase fire preparedness, implementation of community wildfire protection plans, and forest restoration that accomplishes fuel reduction.
Sec. 212. Climate Change Safeguards for Natural Resources Conservation. Establishes an account in Treasury to be called Natural Resources Climate Change Adaptation Account (NRCCAA) to receive proceeds from auction conducted under (new) Sec. 771(b)(7) of Clean Air Act. Funds to be available "without further appropriation" or fiscal year limitation for the purposes of Sec. 370(a)(2), above.	**Sec. 480. Natural Resources Climate Change Adaptation Fund.** See Sec. 480 discussion above.	Senate bill creates one fund; House bill creates two funds, dividing federal and non-federal programs. Both Sec. 212 and Sec. 771(b)(7) in S. 1733, allocate auction proceeds only to Sec.370(a) (2), and not to Sec. 370(a)(3)-(6). If that is intentional, then the source of funding for Sec. 370(a)(3)-(6) is not clear.
Sec. 216. State Programs for Natural Resource Adaptive Activities. Directs Administrator to distribute allowances to states from Sec. 771(a)(15) in accordance with sec. 370(a)(1) (which provides for natural resources adaptation activities; see above).	Sec. 480. See Sec. 480 discussion above.	If Sec. 216 is to be interpreted consistently with Sec. 370(a) and Sec. 771(15)-(16), it appears likely that the reference in this section should be to Sec. 771(a)(16), rather than Sec. 771(a)(15).

S. 1733 (as reported by the Senate EPW Committee)	H.R. 2454 (as passed by the House)	Comments
Sec. 211. Effects of Climate Change on Drinking Water Utilities. Requires EPA, in cooperation with the Secretaries of Commerce, Energy and the Interior, to establish and provide funding for a research program to assist drinking water utilities in adapting to climate change. Research program is to be conducted through a nonprofit research foundation and should address issues related to: water quality and quantity impacts and solutions, impacts on ground water supplies from carbon sequestration, infrastructure impacts, desalination and water reuse, alternative supply technologies, energy efficiency and greenhouse gas minimization, regional cooperative water management solutions, utility management and water management models, improving energy efficiency in water provision and treatment, water conservation and demand management, and customer communication and education. Funding for this program is authorized at $25 million for each of FY2010-2020.	No similar provision	Sec. 211 (of Division A) would establish and fund a broadranging research program that encompasses research in these key issue areas and others. Language similar to Sec. 211 exists in freestanding bills, H.R. 3727 and S.1035. Related bills in this Congress include House-passed H.R. 631, the Water Use Efficiency and Conservation Research Act, which would establish in EPA's Office of Research and Development (R&D) a broad R&D program promoting water use efficiency and conservation to address increasing water scarcity resulting from increased demand and climate change-related effects.
Sec. 381. Water System Mitigation and Adaptation Partnerships. Requires EPA to establish a water system mitigation and adaptation partnership program and to provide grants to states and Indian tribes for water system adaptation projects. Identifies entities eligible to receive project assistance as owners or operators of a community water system, wastewater treatment works, decentralized wastewater treatment system for domestic sewage, groundwater storage and	No similar provision	Sec. 381 authorizes appropriations for water system mitigation and adaptation partnerships. Actual funds would be provided through distribution of emission allowances under Sec. 210 of Division B. Under Sec. 210 of Division B, EPA would distribute the proceeds of emission allowances to states to support the partnership programs under Sec. 381 and a number of other programs. The program proposed by Sec. 381 would consider both adaptation — understanding and planning for impacts on water supplies and

replenishment system, or system for transport and delivery of water for irrigation or conservation. Identifies eligible uses, such as enhancing water use efficiency, modifying or relocating water infrastructure significantly impaired by climate change, or studying how climate change may impact future operations and sustainability of water systems. Provides for a competitive process, prioritizing applications for water systems at the greatest and most immediate risk of facing significant climate-related negative impacts.		watersheds — and mitigation — modifying water infrastructure facilities. Water and wastewater utilities currently are eligible to receive financial assistance for water infrastructure capital projects through Safe Drinking Water Act and Clean Water Act and other federal programs, but these others do not exclusively address climate change-related project needs. S. 1733 does not address coordination between existing infrastructure assistance programs and the proposed Sec. 381 program.
Federal share of projects shall not exceed 50%.		Other legislation in the 111th Congress also addresses water system mitigation and adaptation partnerships. S. 1712 and H.R. 3747 include such a provision (section 6). H.R. 2969 is similar. These bills would direct EPA (not states) to make grants to water systems generally for the same purposes as Sec. 381 of S. 1733.
Sec. 382. Flood Control, Protection, Prevention, and Response. Requires EPA, in consultation with the Army Corps of Engineers and FEMA, to establish a program to provide funds to states and Indian tribes for flood control, protection, prevention, and response projects that address the climate change impacts, with priority to be given to projects that directly assist flood activities by communities, are part of a larger state or watershed plan for flood reduction, advance multiple objectives, protect or enhance natural ecosystem functions, use nonstructural approaches, and reduce the frequency and consequences of flooding in densely populated areas.	No similar provision.	Sec. 382 of S. 1733 authorizes appropriations for a program to grant states and Indian Tribes funds for floodrelated adaptation. Actual funds would be provided through distribution of emission allowances under Division B, Sec. 210. S. 1733 contains language directing EPA to consult with the Corps and FEMA to implement the provision. EPA typically does not undertake flood control activities; however, it does manage grant and loan programs that distribute monies to states and other entities for water quality and other environmental improvement projects. Instead, federal flood damage reduction actions are generally undertaken through the Corps and FEMA.a The relationship of the proposed program to existing federal flood damage reduction activities

		is not defined. S. 1733 promotes a flood risk management approach (e.g., supporting measures that permanently reduce flood risks, such as relocation out of flood-prone areas) and prioritizes opportunities with multiple benefits (e.g., unified flood hazard, built-environment, and ecosystem adaptation measures). A question raised by Sec. 382 is whether eligibility for using these funds would require consistency with state and local hazard mitigation plans and state climate change adaptation plans?
Sec. 384. Coastal and Great Lakes State Adaptation Program. Requires the EPA Administrator to distribute grants for coastal states' (including the Great Lakes states) adaptation. The states may use the funds for planning and addressing the impacts of climate change in coastal watersheds, including (1) addressing sea level and lake level changes, shoreline erosion, and storm frequency or intensity; (2) developing plans for protecting or relocating public facilities and infrastructure; (3) conducting related research and data collection; (4) responding to impacts such as ocean acidification, thermal stratification, saltwater intrusion into freshwater aquifers; algae blooms and species migration. Priority to plan and carry out projects and activities shall be given to state coastal agencies.	No similar provision	Sec. 384 of S. 1733 authorizes appropriations to distribute funds to states for coastal impact adaptation. Actual funds would be provided through distribution of emission allowances under Division B. Sec. 210. This provision would create a mechanism for coastal and Great Lakes states to receive federal grants for climate change adaptation measures. No similarly broad adaptation provision is provided for inland areas; Sec. 382 of S. 1733 (described above) is focused solely on flood-related adaptation. The contents of State Natural Resource Adaptation Plans (SNRAP) outlined in Sec. 369 of Division A are closely related to data and activities identified in Sec. 384. Similarly, there is potential overlap of activities funded in Sec. 384 via state grants and Sec. 370(a)(5) and (6) of Division A for federal activities. Although coordination with other statewide climate change efforts is required in Sec. 384, S. 1733 is neither explicit in the relationship between Sec. 384 and SNRAPs, nor the distinction between the adaptation focus of Sec. 384 and the

	SNRAPs natural resource management focus. Coordinating SNRAPs with existing activities such as State Coastal Zone Management Plans is required by Sec. 369, but coastal activities in Sec. 384 are not subject to the same requirement. One implementation question is whether, and if so how, use of funds and planning under Sec. 384 would be linked to existing state coastal zone management programs and SNRAPs. The relationship between existing National Oceanic and Atmospheric Administration (NOAA) grants under state coastal zone management plans, and EPA's administration of grants under Sec.384 is not defined.

Source: CRS analysis.

a. Other federal agencies are also involved with flood damage reduction projects, such as the United States Department of Agriculture's Natural Resource Conservation Service, the Department of Interior's Bureau of Reclamation and the Tennessee Valley Authority.

Endnotes

[1] For instance, in October, Senators Bingaman, Whitehouse, and Baucus introduced the Natural Resources Climate Adaptation Act of 2009, which would require federal agencies to prepare a national strategy and agency plans to minimize the adverse impacts of climate change on natural resources and maximize resilience.

[2] The Senate Environment and Public Works Committee reported out of committee on November 5, 2009, a revised version of S. 1733, a manager's amendment in the nature of a substitute. S. 1733 was originally introduced by Senators Boxer and Kerry in September 2009. The analysis herein refers to the Senate EPW Committee-reported version of S. 1733, which is available at the Senate EPW Committee website: http://epw.senate.gov/public/index.cfm?FuseAction= Files.View&FileStore_id=1d1bc826-beed-4eb3-933b-d7559bc61d4b.

[3] United States Global Change Research Program, *Global Climate Change Impacts in the United States*, 2009, http://www.globalchange.gov/publications/reports/scientific-assessments/us-impacts

[4] M. L. Parry, O. F. Canziani, J. P. Palutikof, P. J. van der Linden, and C. E. Hanson, eds., *Climate Change 2007: Impacts, Adaptation and Vulnerability*, Contribution of Working Group II to the Fourth Assessment Report of the Intergovernmental Panel on Climate Change, Cambridge, UK.

[5] National Research Council, *Restructuring Federal Climate Research to Meet the Challenges of Climate Change,* 2009, http://www.nap.edu/catalog.php?record_id=12595.

[6] House Select Committee on Energy Independence and Global Warming, "Building U.S. Resilience to Global Warming Impacts," hearing held on October 22, 2009, http://globalwarming.house.gov/pubs?id =0011#main_content.

[7] General Accountability Office (GAO), *Climate Change Adaptation—Strategic Federal Planning Could Help Government Officials Make More Informed Decisions*, GAO-10-1 13, October 2009, http://www.gao.gov/ new.items/ d10113.pdf.

[8] For an overview of what 13 federal agencies are doing related to climate change adaptation, see Government Accountability Office (GAO), *Climate Change Adaptation: Information on Selected Federal Efforts to Adapt to a Changing Climate*, GAO-10-1 14SP, October 7, 2009, an E-supplement to GAO-10-1 13, http://www.gao.gov/ new.items/d101 14sp.pdf.

[9] See http://www.doi.gov/news/09_News_Releases/091409.html.

[10] Secretarial Order No. 3289, http://www.doi.gov/climatechange/SecOrder3289.pdf.

[11] See http://www.energy.ca.gov/2009publications/CNRA-1000-2009-027/CNRA-1000-2009-027-D.PDF.

[12] In a cap-and-trade system, companies or other groups are issued a number of emission *allowances* (or *credits*) which represent the right to emit a specific amount of greenhouse gases. The total amount of allowances and credits cannot exceed the cap, limiting total emissions to that level. Policymakers decide how, to whom, and for what purpose to distribute emission allowances. The allowances represent significant value in terms of a wealth transfer in the case of directly allocated allowances or government revenue in the case of auctioned allowances. For more information, see CRS Report RL34502, *Emission Allowance Allocation in a Cap-and-Trade Program: Options and Considerations*, by Jonathan L. Ramseur.

[13] For a comparison of cap-and-trade provisions, including a discussion of allowance allocation differences between S. 1733 and H.R. 2454, see CRS Report R40896, *Climate Change: Comparison of the Cap-and-Trade Provisions in H.R. 2454 and S. 1733*, by Brent D. Yacobucci, Jonathan L. Ramseur, and Larry Parker.

[14] See Sec. 211, Sec. 212, and Sec. 370(a)(2)-(6) in **Appendix**.

[15] See H.Rept. 111-137, Table 4, pp. 379-380 .

[16] See http://www.ipcc.ch/publications_and_data/publications_ipcc_fourth_assessment_report_wg2_report_impacts_adaptation_and_vulnerability.htm.

[17] Martin Parry et al., *Assessing the Costs of Adaptation to Climate Change: A Review of the UNFCCC and Other Recent Estimates* (London: International Institute for Environment and Development (IIED), August 2009), http://74.125.93.132/search?q=cache:KCCoQ47xQdMJ:www.iied.org/pubs/pdfs/11501IIED.pdf+%22 Assessing+the+costs+of+adaptation%22&cd=2&hl=en&ct=clnk&gl=us&client=firefox-a.

[18] World Bank, *Economics of Adaptation to Climate Change: New Methods and Estimates* (consultation draft), September 2009, http://beta.worldbank.org/climatechange/content/economics-adaptation-

[19] BBC News, *EU Accelerates Climate Funding*, http://news.bbc.co.uk/2/hi/europe/8334146.stm.

[20] This Board would also oversee programs to promote "supplemental emission reductions," greenhouse gas sequestration in forests under Sec. 751.

[21] H.R. 2454, Part 1, would repeal and replace parts of the existing Global Change Research Act of 1990. This provides for the continuation and coordination of federal global change research. The U.S. Global Change Research Program (USGCRP) established under the GCRA of 1990 and continued under both bills has been the primary vehicle in the United States for domestic and internationally coordinated research on climate change. It has produced world-leading results in many aspects of climate change science. However, it also has been criticized for not being sufficiently oriented toward the information needs of potential users, especially decision-makers, as well as for insufficient interagency coordination and budget prioritization. For more information, see CRS Report RL338 17, *Climate Change: Federal Program Funding and Tax Incentives*, by Jane A. Leggett.

[22] H.R. 2454 uses "climate adaptation" in these sections, raising the question of whether plans and funded activities are to include climate variability as well as climate change.

[23] For example, property rights were a major issue in certain conservation bills in the 106th Congress; see out-of-print CRS Report RL3 0444, *Conservation and Reinvestment Act (CARA) (H.R. 701) and a Related Initiative in the 106th Congress*, by Jeffrey A. Zinn and M. Lynne Corn (available from Lynne Corn).

[24] For a discussion of the current status of geospatial research and coordination efforts, see CRS Report R40625, *Geospatial Information and Geographic Information Systems (GIS): Current Issues and Future Challenges*, by Peter Folger.

[25] Not only are extreme events a concern, but also of concern are anticipated changes in average streamflows, groundwater recharge rates, and timing and depth of snowpack. Ocean, coastal, and marine adaptation issues are generally not discussed in this section, except as they relate to § 384 and flooding and shoreline protection.

[26] Intergovernmental Panel on Climate Change, *Climate Change 2007: Impacts, Adaptation and Vulnerability*, Chapter 3, Freshwater Resources and Their Management, Contribution of Working Group II to the Fourth Assessment Report, Cambridge, UK, 2007, http://www.ipcc.ch/publications_and_data/publications_ipcc_fourth_assessment_report_wg2_report_impacts_adaptation_and_vulnerability.htm.

[27] Water-related functions are shared by all levels of government and the private sector. Local governments and other public and private entities (e.g., water utilities) are largely responsible for municipal water infrastructure (e.g., drinking water, wastewater, stormwater) and flood damage reduction measures. The states generally allocate water within their jurisdiction. The federal government generally participates in water projects that are considered to be in the national interest (e.g., navigation to support commerce, dams and related irrigation to promote settlement of western states, participation in the construction of congressionally authorized flood protection projects). Federal water activities are spread over numerous federal agencies.

[28] Both S. 1733 (as reported by the Senate EPW Committee) and H.R. 2454 also include other water-specific provisions, which are not discussed herein because they are addressed in the bill not as part of adaptation, but through greenhouse gas reduction programs. These provisions are focused on the energy efficiency gained by improving water efficiency, and include §§ 141-143 of S. 1733 and §§ 215-217 of H.R. 2454. Also, § 157 of S. 1733, which would require a study of risk-based policies and programs (including flood insurance), is related to water-related adaptation. Other bills, such as S. 1462, also address (primarily through studies) water-related issues that may arise related to the energy sector, including water use of lower carbon dioxide-emitting electricity technologies.

[29] Natural resources adaptation funding is discussed in the previous section. Whether the funds made available by these bills for aquatic ecosystem restoration would cover the anticipated cost of adapting to climate change is uncertain because reliable estimates of these costs are not available.

[30] Reliable estimates of the federal water resources infrastructure costs associated with adaptation are not available.

CHAPTER SOURCES

The following chapters have been previously published:

Chapter 1 – This is an edited, excerpted and augmented edition of a United States Government Accountability Office publication, Report Order Code GAO-10-113, dated October 2009.

Chapter 2 – These remarks were delivered as Statement of Edward J. Markey before the Select Committee on Energy Independence and Global Warming, given October 22, 2009.

Chapter 3 – This is an edited, excerpted and augmented edition of a United States Government Accountability Office publication, Report Order Code GAO-10-175T, dated October 22, 2009.

Chapter 4 – These remarks were delivered as Statement of Eric Schwaab before the Select Committee on Energy Independence and Global Warming, given October 22, 2009.

Chapter 5 – These remarks were delivered as Statement of Stephen Seidel before the Select Committee on Energy Independence and Global Warming, given October 22, 2009.

Chapter 6 – These remarks were delivered as Statement of Dr. Kenneth P. Green before the Select Committee on Energy Independence and Global Warming, given October 22, 2009.

Chapter 7 – This is an edited, excerpted and augmented edition of a United States Congressional Research Service publication, Report Order Code R40911, dated November 12, 2009.

INDEX

A

accountability, 54, 85
Afghanistan, 73
Africa, 13, 73, 122
agriculture, 12, 21, 91, 114, 116, 117, 118, 120, 121, 132
Air Force, 12, 61, 100
air quality, 116, 127, 138
Alaska, 76, 128, 138
algae, 151
alternative energy, 130
appropriations, 114, 120, 129, 131, 145, 146, 149, 150, 151
aquifers, 151
architects, 94
armed forces, 12
Asia, 13, 73
assessment, 10, 17, 19, 24, 45, 52, 117, 125, 131, 135, 136, 141, 154
assets, 100, 101, 102, 117
authorities, vii, ix, 1, 2, 6, 9, 14, 24, 36, 38, 55, 57, 82, 83, 125, 126, 127

B

background, 6, 57, 58, 109
background information, 57
barriers, 36, 53, 85, 93, 102
biodiversity, 10, 13
birds, 73, 145
building blocks, 13, 83, 103
Bureau of Land Management, 48, 73

C

capital projects, 149
carbon, 15, 72, 80, 86, 94, 95, 99, 120, 130, 138, 140, 149, 155
carbon dioxide, 72, 86, 99, 155
carbon emissions, 15
case study, 24, 104
challenges, vii, viii, ix, 1, 2, 3, 4, 6, 9, 11, 24, 27, 28, 30, 31, 34, 35, 36, 38, 40, 42, 45, 53, 54, 55, 56, 58, 59, 76, 81, 82, 84, 85, 97, 102, 112, 116, 117, 130, 131, 132
China, 110
City, 2, 6, 7, 14, 15, 16, 17, 27, 31, 41, 52, 57, 72, 73, 76, 82, 83, 100
civil society, 14
Clean Air Act, 39, 148
clean energy, 80
Climate Change Science Program, vi, 10, 56, 72, 73, 91, 95
CO_2, 72, 86, 94
Coast Guard, 12
coastal communities, 9, 41, 93, 118
communication, 17, 41, 128, 149
community, ix, 12, 97, 104, 148, 149
compatibility, 147
competitive process, 149
complexity, 31, 54, 45, 46, 75, 85
compliance, 127, 140
composition, 91
computer systems, 54
conference, 18, 73
conflict, 54, 76

Congressional Budget Office, 121
conservation, 22, 42, 73, 92, 112, 131, 140, 143, 145, 147, 149, 154
consolidation, 48
consulting, 133
cooling, 99
coordination, 7, 13, 17, 36, 86, 92, 93, 102, 118, 130, 135, 136, 147, 149, 151, 154
cost, 4, 50, 100, 111, 112, 130, 136, 140, 148, 155
cost of living, 111, 112
cost saving, 136
critical infrastructure, 11, 15
crop production, 122
crop rotations, 101
crops, viii, 39, 80

D

data analysis, 104
data collection, 9, 151
database, 13, 59
decision makers, 2, 8, 9, 10, 14, 17, 22, 30, 32, 35, 45, 46, 47, 48, 54, 49, 83, 135
decision-making process, 35, 49
dengue, 122
Department of Agriculture, 39, 60, 72, 104, 132, 152
Department of Commerce, 61, 72, 126, 130
Department of Defense, 10, 12, 61, 72, 100
Department of Energy, 9, 10, 11, 61, 132
Department of Health and Human Services, 61
Department of Homeland Security, 11, 56, 61, 77
Department of the Interior, 8, 61, 83
deposits, 121, 145
developed countries, 23
developing countries, 8, 12, 110, 122, 133, 134
developing nations, 122
direct cost, 130
disaster, viii, 53, 79, 136, 140
disclosure, 133, 147
dissolved oxygen, 131
diversity, 10

draft, 9, 11, 14, 59, 72, 118, 154
drinking water, x, 112, 114, 119, 131, 149, 154
drought, 4, 18, 25, 39, 94, 99, 110, 130, 131, 142, 144

E

ecological restoration, 140
economic growth, 112
economic losses, 99
economic well-being, 90
economy, 80, 99, 100, 115, 118
ecosystem, 17, 40, 73, 94, 103, 131, 132, 147, 150, 155
ecosystem restoration, 131, 132, 155
Education, 40, 51, 60, 147
educational process, 41
elaboration, 140
electricity, 99, 111, 155
elephants, 73
eligibility criteria, 133
emergency response, 136, 140
emission, x, 114, 119, 120, 125, 126, 127, 128, 129, 134, 139, 141, 145, 149, 150, 151, 153, 154
encouragement, 145
endangered species, 145, 147
energy efficiency, 115, 149, 155
enforcement, 140
engineering, 19
England, 25, 26, 27
environmental conditions, 93
environmental degradation, 140
environmental impact, 106
environmental issues, 44
environmental policy, 109
environmental protection, 25
Environmental Protection Agency, 8, 56, 61, 72, 73, 127, 132
EPA, 8, 11, 12, 72, 103, 106, 127, 134, 136, 138, 140, 141, 143, 147, 149, 150, 151
equipment, 16, 148
erosion, ix, 21, 22, 24, 72, 90, 151
EU, 110, 122, 154
European Union, 25

Everglades, 100
Executive Order, ix, 9, 72, 74, 89, 91, 95, 103
executive orders, 14, 18, 83
expertise, 6, 47, 48, 59, 101, 124, 133
experts, 51, 99, 104, 117
exposure, 39, 109

F

farmers, 117
farms, viii, 79
federal authorities, 55
federal law, 39, 52
FEMA, 11, 38, 75, 132, 143, 150
financial resources, 127
financial support, 17, 93, 142
fish, 91, 92, 94, 147
Fish and Wildlife Service, 10, 103, 145
flexibility, 51
flood hazards, 101
flooding, ix, 15, 16, 17, 18, 19, 20, 24, 31, 46, 52, 53, 90, 99, 100, 111, 112, 132, 142, 144, 150, 154
forecasting, x, 110
forest ecosystem, 94
forest management, 94
forest restoration, 148
freshwater, 10, 91, 111, 147, 151
funding, 15, 24, 25, 28, 7, 30, 49, 50, 51, 44, 50, 91, 93, 94, 118, 121, 122, 123, 124, 125, 126, 128, 129, 131, 132, 133, 138, 139, 141, 145, 146, 147, 148, 149, 155

G

Global Change Research Act, 52, 135, 154
grant programs, 121
grasses, 90
grasslands, 10, 146
Great Lakes, x, 114, 119, 131, 151
greenhouse gas emissions, 4, 98, 110, 115
greenhouse gases, vii, viii, ix, x, 1, 3, 4, 72, 81, 86, 97, 98, 110, 112, 153
groundwater, 149, 154

guidance, 10, 11, 12, 13, 22, 23, 27, 35, 38, 48, 49, 50, 52, 47, 50, 72, 106, 133
Gulf Coast, 11, 72

H

habitats, 93, 129
hazards, 11, 22, 23, 52, 74, 92, 130
health effects, 11, 12, 128
heat stroke, 122
Hurricane Katrina, viii, 17, 79
hurricanes, 1, 14, 52, 83, 116, 117

I

impact assessment, 11, 141
Independence, v, vi, 3, 79, 81, 89, 97, 109, 117, 153
India, 110
industrial sectors, 116
industrialized countries, 122
infestations, 101
information sharing, 9, 25, 47, 49, 103
interagency coordination, 126, 135, 154
interest groups, 119
invasion of privacy, 147
issues, vii, viii, 1, 3, 6, 10, 32, 39, 41, 39, 44, 53, 54, 55, 57, 48, 82, 85, 91, 92, 122, 132, 149, 154, 155

J

jurisdiction, 48, 55, 86, 131, 143, 144, 154

L

land acquisition, 129, 146
landscape, 10, 94
landscapes, 4, 9
Latin America, 13
leadership, viii, 18, 28, 30, 54, 79, 85, 86, 95, 98, 101, 103, 105, 118
learning, 41, 46
legislation, x, 6, 14, 24, 83, 90, 105, 110, 114, 120, 131, 150
levees, viii, 18, 79, 131
livestock, 116

local authorities, 55
local government, 8, 14, 17, 23, 25, 35, 36, 38, 39, 44, 48, 49, 52, 53, 55, 80, 85, 86, 92, 93, 95, 100, 101, 102, 121, 127, 137, 138, 139, 143
lying, ix, 20, 90

M

major decisions, 102
management, viii, 4, 6, 9, 11, 13, 14, 19, 30, 31, 34, 38, 39, 40, 50, 52, 54, 46, 50, 72, 74, 76, 81, 83, 91, 102, 103, 116, 117, 130, 131, 135, 140, 145, 147, 149, 151
mandates, 36, 111
mapping, 13, 22, 130, 148
Marine Corps, 12, 61
marketing, 132
medication, 80
melt, viii, 79, 122, 135
melting, viii, 79, 122
membership, 41
messages, 59
meter, 100
methodology, 7, 82
Mexico, 73
Middle East, 73
migration, 22, 151
missions, 36, 53

N

National Aeronautics and Space Administration, 8, 61
National Defense Authorization Act, 12, 73
National Institutes of Health, 12, 61
National Park Service, 73
national parks, 100
National Research Council, viii, 4, 73, 75, 76, 81, 117, 141, 153
National Science Foundation, 8
national security, 8, 12, 80, 99, 134
national strategy, 92, 93, 128, 153
native species, 4
natural disasters, 12, 14, 15, 52, 83

natural resource management, 8, 9, 72, 83, 151
natural resources, x, 3, 6, 9, 72, 83, 90, 92, 93, 99, 104, 114, 118, 119, 120, 125, 127, 128, 129, 131, 132, 140, 142, 143, 144, 145, 147, 148, 153
NCS, 126, 137
Nepal, 23
New Zealand, 73
nitrous oxide, 72, 86

O

oceans, viii, 75, 79, 98
opportunities, viii, 4, 10, 12, 23, 81, 100, 117, 129, 134, 143, 146, 150
outreach, 104, 135
oversight, 15, 124, 129

P

Pacific, 1, 10
Parliament, 24
per capita income, 138
performance, 7, 25, 7, 46, 59, 44, 134
personal responsibility, ix, 109
pests, viii, 79, 116
physical interaction, 142
planning decisions, 27, 32, 41, 53
plants, 15, 16, 21, 99, 112, 145
policy makers, 147
policy options, 10, 41, 42, 39, 44, 49, 48, 75
pollution, 80, 100, 112, 130, 138
population growth, 15, 116
poverty, 110
precipitation, 1, 31, 34, 45, 91, 94, 99, 101, 104, 124, 130
predictability, 91
preparedness, 12, 53, 128, 148
prevention, x, 114, 119, 138, 150
price signals, 112
private investment, 90
private ownership, 94
project, 10, 12, 16, 27, 52, 72, 132, 138, 149
properties, 75
property rights, 129, 146, 154

Index

public awareness, 14, 7, 30, 44, 83
public education, 41
public health, x, 7, 8, 11, 12, 58, 106, 114, 119, 125, 127, 128, 141
public investment, 92
public safety, 90

R

rainfall, 19, 117
rangeland, 48
reactions, 31
reading, x, 110
reality, 32, 55, 99
recognition, 4, 15, 23, 39, 84, 99, 101
recommendations, iv, 2, 12, 75, 102, 117, 134
recreation, 17, 94
reliability, 112, 130, 138
remote sensing, 73
renewable energy, 94, 99, 115
repair, 18
requirements, 122, 123, 125, 133, 137, 140
resilience, 9, 18, 24, 25, 80, 83, 102, 110, 112, 129, 138, 140, 142, 153
resistance, 54, 85
resource availability, 131
resource management, 25, 30, 38, 40, 48, 52, 73, 92, 118, 135
resources, 2, 4, 9, 10, 14, 22, 27, 28, 7, 31, 34, 35, 38, 48, 52, 54, 55, 44, 83, 84, 85, 86, 91, 92, 93, 99, 101, 102, 116, 118, 119, 123, 131, 132, 133, 142, 144, 145, 147, 155
revenue, 93, 111, 112, 114, 121, 128, 141, 153
risk assessment, 24, 25
risk management, 39, 150
risk-taking, 110, 111
rodents, 128
role-playing, 41
runoff, 19, 20, 112, 130, 131

S

salmon, 2, 9, 34, 84

sample survey, 58
sampling error, 58
SAP, 72, 73, 75
scarcity, viii, 79, 99, 149
scientific understanding, ix, 93, 98
sea level, vii, viii, ix, 1, 2, 3, 4, 10, 11, 14, 15, 16, 18, 19, 20, 21, 22, 23, 24, 26, 39, 44, 46, 52, 72, 79, 81, 89, 90, 91, 92, 98, 100, 101, 104, 117, 130, 132, 135, 151
sea-level, 116, 118
sea-level rise, 116, 118
Secretary of Commerce, 137, 147
Secretary of Defense, 61
Senate, x, 72, 113, 114, 115, 119, 120, 121, 123, 124, 125, 127, 129, 130, 133, 137, 138, 141, 142, 143, 144, 145, 146, 147, 148, 149, 153, 155
sensitivity, 26, 115
settlements, 90, 94, 116
sewage, 149
shelter, viii, 79
shoreline, ix, 21, 22, 74, 90, 101, 117, 151, 154
smart com, 41
snaps, 112
soil erosion, 130
species, 23, 34, 35, 39, 73, 91, 94, 122, 128, 129, 131, 151
stabilization, 21
stakeholders, 6, 7, 46, 57, 82, 102, 104, 125, 133, 137, 141, 143
storage, 130, 149
storms, viii, 2, 1, 14, 18, 19, 20, 21, 25, 46, 79, 90, 91, 99, 110
stormwater, 17, 19, 154
Strategic Petroleum Reserve, 11
strategic planning, vii, 3, 53, 85, 92, 103, 105, 118
strategy, 2, 3, 4, 14, 21, 50, 72, 80, 85, 91, 92, 94, 103, 118, 124, 125, 128, 135, 142, 143
stratification, 151
substitution, x, 113
sulfur, 72, 86
supervision, 129, 146

T

technical assistance, 13, 20, 22, 144, 146
temperature, 3, 1, 9, 31, 34, 45, 91, 94, 99, 104, 117, 122, 124
tension, 40
tensions, 99
testing, 124
threats, 116, 128, 130
tides, 74, 129
time frame, 30, 136
Title I, 135
Title II, 135
training, 3, 13, 27, 40, 41, 44, 51, 85, 100, 109, 124, 127, 148
training programs, 41, 148
trajectory, 110
transportation, 11, 17, 18, 91, 100, 104, 111, 114, 118, 120, 121, 129, 138
transportation infrastructure, 129
tropical storms, 3

U

U.S. Geological Survey, 11
UK, 27, 74, 153, 154
United Kingdom, 2, 6, 8, 24, 25, 26, 27, 32, 46, 57, 74, 82, 84
United Nations, 23, 112, 122, 133
United States, 157
universities, 14, 22, 51, 83
USDA, 39, 51, 76, 130, 143

V

vegetation, 19, 20, 117, 129
visualization, 13
vulnerability, ix, 2, 3, 4, 10, 11, 15, 20, 23, 45, 47, 49, 85, 90, 92, 117, 124, 125, 127, 135, 136, 154

W

Wales, 25
waste, 17, 18, 54
wastewater, 15, 17, 18, 45, 46, 111, 112, 130, 149, 154
water quality, 91, 131, 138, 149, 150
water resources, 11, 39, 91, 103, 116, 118, 130, 132, 147, 155
water supplies, 132, 149
water vapor, 75
watershed, 14, 17, 150
wealth, 38, 153
wetlands, 20, 72, 74, 90, 93, 94, 99, 145
White House, 103, 126, 135
wildfire, x, 114, 119, 130, 138, 142, 144, 148
wildland, 73
wildlife, 9, 92, 94, 129, 145, 147
witnesses, 80
working groups, 58, 104
World Bank, 122, 154